La Fayette Coll. Feb 23. 53

Dear Sir

I owe an apology for having so long delayed to send you a copy of my treatise on Conics &c, after the reception of your favor of last Month. I thought I had done so, till happening to look over my list today, I found that I had not. It is possible that I delayed, thinking that I should meet you at Cleveland last August at the meeting of the American Association, which it was proposed to hold there at that time. I now send you one by the same mail as this & hope it will reach you safely.

Hoping also that I shall have the pleasure of seeing you at Cleveland this year, I remain

Yours sincerely

Jas. H Coffin

Prof Geo Denison

ELEMENTS

OF

CONIC SECTIONS

AND

ANALYTICAL GEOMETRY.

BY

JAMES H. COFFIN, A. M.

PROFESSOR OF MATHEMATICS AND PHYSICS IN LAFAYETTE COLLEGE, AND
AUTHOR OF A TREATISE ON SOLAR AND LUNAR ECLIPSES,
ASTRONOMICAL TABLES, ETC.

REVISED AND STEREOTYPED.

NEW YORK:
PUBLISHED BY ROBERT B. COLLINS,
NO. 254 PEARL-STREET.
1851.

Entered according to Act of Congress, in the year 1848,

By JAMES H. COFFIN,

In the Clerk's Office of the Eastern District of Pennsylvania.

PREFACE.

THE following treatise has been prepared to meet the wants of the author, in the instruction of his classes. He has felt the need of a work on the Conic Sections, that was not, on the one hand, so prolix and tedious in the method of demonstration as to render the study repulsive to the student; nor, on the other, so meager as to the number of properties discussed, as to give him but a very imperfect idea of the interesting features of these curves, and materially to cripple his future course of study, which, if properly conducted, requires a thorough knowledge of them. In the preparation of this work, it has been the aim to avoid both these defects: so as, on the one hand, to render it as full and complete as the most thorough works in use upon the subject; and, on the other, to lighten the labor of the student, by simplifying the demonstrations without rendering them less rigid,—thus giving him a more clear and perfect knowledge of the properties discussed, and at the same time diminishing the size of the book.

The properties of the Conic Sections may be investigated by either of two quite dissimilar methods; each of which has its peculiar advantages. We may study them directly from the figure itself, in the same manner as in elementary geometry; and this method, which is called the *geometrical,* has the advantage of affording a more clear conception of the properties under consideration. Or we may, after the method invented by Descartes, first represent the several parts of the figure by an equation, and then proceed in our investigations by pure algebra. This method, which is called the *analytical*,[a] has the advantage of enabling us to extend our researches far beyond what we could otherwise do; just as, by the aid of ordinary algebra, we can solve questions which it would be impossible to solve by common arithmetic. The former method is better adapted to make clear and sound reasoners; the latter, when used in its proper sphere, expert and finished mathematicians. It cannot, however, profitably supersede the geometrical method in cases to which the latter is applicable. Indeed, as just hinted, the analytical method is to geometry what algebra is to common arithmetic—valuable as an aid, but absurd as a substitute. It has been sometimes supposed that the properties of the Conic Sections could be more easily investigated by the analytical method, and the exceedingly tedious geometrical demonstrations that we find in some works, certainly afford grounds for the opinion. But it need not be so. All the leading properties can be *demonstrated* with equal ease, and greater clearness, by the geometrical method, while it is the province of analytical geometry to *apply* them. A knowledge of both is, therefore, essential to a perfect course in mathematics.

In accordance with the foregoing views, this treatise consists of two parts. In the First Part the various properties of the Conic Sections are demonstrated, for the most part *geometrically;* and, in the Second, the student is taught how to represent lines, curves, and surfaces *analytically*, and to solve problems relating to them.

---

[a] Often called the French method.

Our definition of a Conic Section, is merely an extension of the common definition of a parabola, and is recommended by the following considerations:—

I. *It is general, belonging to each of the three curves.* The more common method is, to define the ellipse, parabola, and hyperbola as three distinct curves. They are called conic sections, but is not the student left in the dark as to what a conic section is, or why these curves are called by the same general name?

II. By thus uniting the sections, and showing that instead of being three different curves, they are merely modifications of one and the same curve—the conic section, *the mind of the student is better prepared to appreciate the analogies that he finds between them.*

III. *It simplifies the demonstrations,* as it enables us at the outset to prove both of the fundamental propositions of other treatises independently of each other; so that we can avail ourselves of either at pleasure in the subsequent demonstrations. It is to this fact chiefly that many of the demonstrations owe their simplicity.

With a view to keep the analogies between the three curves prominent before the mind of the student, it is the author's practice with his classes to take up the corresponding propositions in connection, instead of following the order of the book; and for the sake of convenience in giving out the lessons, they are numbered alike in the second, third, and fourth chapters.

The subject of the *curvature* of the Conic Sections can be discussed to better advantage by the aid of the Differential Calculus; but for the benefit of those who are not acquainted with that branch of mathematics, Chapter V. is given as a substitute. The last proposition in this chapter, and the last two in Chapter VI., discuss properties not treated of in other works on the Conic Sections, but thought to be important from their applications in physical astronomy.

In the preparation of the Second Part, special pains have been taken to present the subjects to be discussed in such a light as to secure the two following objects, viz.:—1st. That the student, when he commences a proposition, may see definitely what he wants to accomplish; and 2d. That the algebraic results at which he arrives may admit of easy interpretation, and the identity between them and the geometrical relations which they represent, readily and clearly seen. For several of the problems, as also in some degree for the plan, the author is indebted to the valuable treatise on the Calculus by Professor M'Cartney, to which this work is intended as an introduction.

A Key for the use of Teachers, containing solutions of the more difficult problems in Part II., is published separately.

CONTENTS.

CONIC SECTIONS.

ANALYTICAL GEOMETRY.

PART I.

CONIC SECTIONS.

CHAPTER I.

DEFINITIONS AND GENERAL PROPOSITIONS.

(1) *A conic section is a curve, the distance of any point in which from a given point, is to its distance from a given straight line in a given ratio.*

If the distance to the point be less *than to the line, the curve is called an Ellipse; if* equal, *a Parabola; and if* greater, *an Hyperbola.*[a]

For example; in the following figure, in which AB represents the given line, F the given point, and R any point in the curve MN, if the ratio of RS to RF continues the same wherever in the curve the point R is taken, the curve is a conic section, and is an ellipse, parabola, or hyperbola, according as RF is less, equal to, or greater than RS.

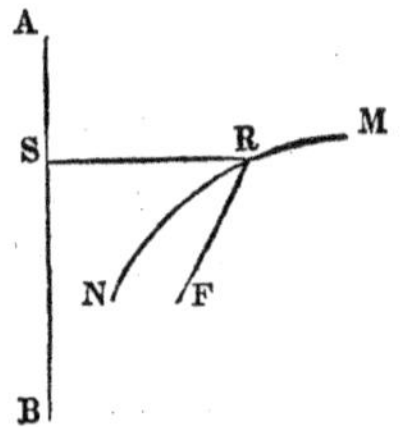

[a] If the distance to the line be infinite, the curve becomes a circle; and if the distance to the point be infinite, the curve becomes a straight line.

(2) Prop. I. Problem.

To describe a conic section.

(Fig. 1. Ellipse.)

For Parabola and Hyperbola see page 153.

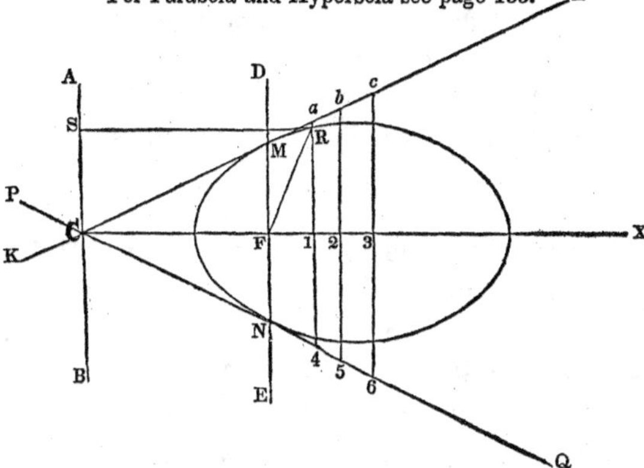

Let F be the given point, AB the given straight line, and $m : n$ the given ratio.

Through F draw DE parallel, and CX perpendicular to AB, each of indefinite length. From FD and FE cut off FM and FN, so that each shall be to FC in the given ratio; that is, FM or FN : FC :: m : n. Through the points C.M and C.N draw KL and PQ of indefinite length, and from them draw a series of perpendiculars to CX, as a.1, b.2, c.3, 4.1, 5.2, 6.3, &c. Take the length of any perpendicular in the compasses, and with one foot on F, note where the other falls on that perpendicular. The points thus found will indicate the curve.

Let R be a point found as above described. Join FR, and draw RS perpendicular to AB.

By construction	MF : CF :: m : n.
But	MF : CF :: a.1=RF : C.1=RS.
Therefore	RF : RS :: m : n.

In the same manner it may be shown, that the distance of any other point in the curve from F is to its distance from AB in the given ratio; and hence the curve is a conic section.

(3) The given point F is called the *focus*, and the given straight line AB the *directrix*.

(4) The portion of CX intercepted between its intersections with the curve, is called the *transverse* or *major axis*.

(5) The middle point, as C, (Fig. 2,) of the transverse axis is called the *centre* of the curve, and its extremities, A and B, *vertices*.

(6) The distances from the focus to the vertices, FA and FB, are called *focal distances*.

(7) The *conjugate axis*, DE, is a straight line drawn through the centre, at right angles to the transverse axis, bisected by it, and equal to twice the mean proportional between the focal distances. Since the parabola intersects CX in only one point, its transverse axis is infinite, and it has no conjugate axis nor centre.

(8) Any straight line drawn through the centre, and limited both ways by the curve, is called a *diameter*, and its extremities its vertices; as HI (Fig. 6.)

(9) Two diameters are said to be *conjugate*, when each is parallel to a tangent to the curve at the extremity of the other; as NU and PL (Fig. 10.) [*See Appendix*, E.]

(10) An *ordinate* to any diameter is a straight line parallel to a tangent at its vertex, and limited in one direction by the curve, and in the other by the diameter; as RV (Fig. 2,) or DZ (Fig. 14.) If produced, so as to be limited in both directions by the curve, it is called a *double ordinate;* as RU (Fig. 2.)

(11) The portions into which an ordinate divides a diameter, are called *abscissas*.

(12) The parameter of any diameter is the third proportional to i and its conjugate. In the parabola it is the third proportional to any abscissa and its corresponding ordinate. The parameter of the transverse axis is called the *principal parameter*, or *latus-rectum*.

(13) The lines KL and PQ are called *focal tangents*.

(14) The distance from the focus to the centre is called the eccentricity.

(15) A line drawn perpendicular to a tangent to the curve, from the point of contact, is called a *normal line;* and the part of the transverse axis intercepted between it and an ordinate let fall from the point of contact, is called a *subnormal*.

(15*a*) *Cor.* The distance from any point in the curve to the focus is equal to a perpendicular to the transverse axis, drawn through the point from the focal tangent.

(16) Prop. II. Problem.

To describe a conic section that shall pass through three given points, and have a given focus.

Let M, N, and P be the three given points, and F the given focus.

Join FM, FN, FP, MN, and NP, and produce MN and NP to R and L, making

MR : NR :: MF : NF,
and NL : PL :: NF : PF.

Through the points R and L draw the line ST of indefinite length, and perpendicular to it draw PI, NH, MG, and FK.

(Fig. 1*a*.)

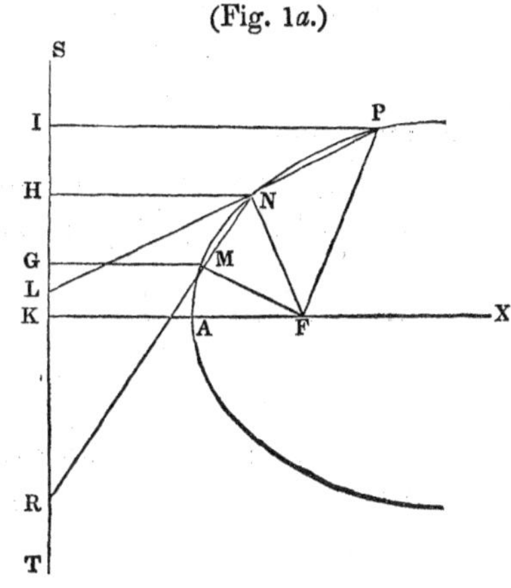

By sim. tri.	GM : HN :: MR : NR.
But, by construction,	MF : NF :: MR : NR.
Therefore, alternately,	GM : MF :: HN : NF.
Also sim. tri.	IP : HN :: PL : NL.
And by construction,	PF : NF :: PL : NL.
Therefore, alternately,	IP : PF :: HN : NF.

Thus the distance of each of the points M, N, and P, from the line ST, is to its distance from F in the same ratio; and, consequently, if with this ratio we describe a conic section, in the same manner as in Prop. I., making F the focus and ST the directrix, it will pass through the points M, N, and P.

This proposition is important, as it enables us to determine the orbit of a planet or comet by means of three observations.[a]

(16*a*) *Scholium.* The conic sections may be formed by the intersection of a plane with the sides of a cone, and hence their name. If the cutting plane be parallel to one of the sides of the cone, the curve is a parabola; if more nearly perpendicular to the axis of the cone, it is an ellipse; and if less so, an hyperbola. If quite perpendicular, the section is evidently a circle. And, universally, the ratio mentioned in (1) is the ratio of the sines of the angles, which the cutting plane and the sides of the cone form with the base. See Appendix, Note A.

---

[a] Bridge's Conic Sections.

CHAPTER II.

OF THE ELLIPSE.

(17) Prop. I. Theorem.

The squares of ordinates to the transverse axis of an ellipse are to each other as the rectangles of the corresponding abscissas.

That is, $GF^2 : RV^2 :: AF.FB : AV.VB$.

In which GF is the focal ordinate, and RV any other ordinate.

Through the vertices A and B draw, from the focal tangent IT, the lines LM and IK at right angles to AB; through the focus F draw LF and IF, meeting LM and IK in M and K; and produce RV both ways to P and N.

(Fig. 2.)

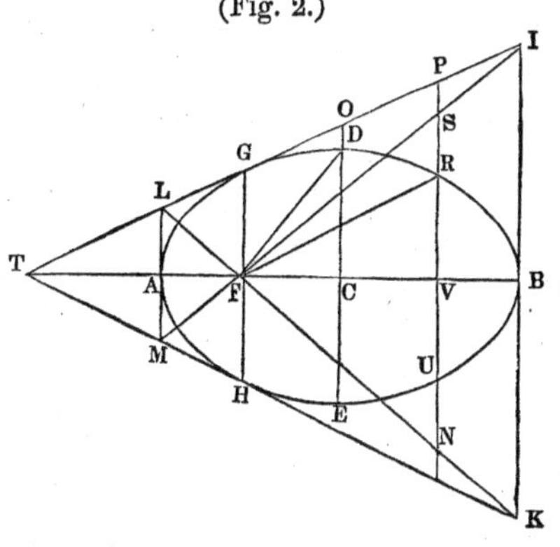

By the principles of construction (15[a]),

$$IB=BF, \text{ and } AL=AF.$$

But[a] $IB : BF :: SV : FV$, and $AL : AF :: VN : FV$.

Therefore $SV=FV$, and $VN=FV$.

By similar triangles (LGF and LPN) we have

$$GF : PN :: LG : LP :: {}^{b}AF : AV,$$

and by similar triangles (GIF and PIS) we have

$$GF : PS :: GI : PI :: {}^{b}FB : VB.$$

---

[a] Leg. 4. 18. Euc. 6. 4. [b] Leg. 4. 15, Cor. 2. Euc. 6. 2.

Multiplying the proportions together,

$$GF^2 : PN.PS :: AF.FB : AV.VB.$$

But $PS = PV - SV =$ (by 2) $FR - SV = FR - FV$;

and $PN = PV + VN =$ (by 2) $FR + VN = FR + FV$.

Therefore $PN.PS = \overline{FR+FV}.\overline{FR-FV} =$ [a] $FR^2 - FV^2 =$ [b] RV^2.

Substituting this value of PN.PS, we have

$$GF^2 : RV^2 :: AF.FB : AV.VB.$$

(18) *Cor.* 1. Hence, if two ordinates are equally distant from the centre, they are equal to one another.

That is, if $CF = CV$, then $GF = RV$.

(19) *Cor.* 2. Hence, the two portions of the curve lying on the opposite sides of the transverse axis AB, or the conjugate axis DE, are symmetrical; and if placed upon one another, would coincide in every part. For if at any point they should not coincide, the ordinates at that point would be unequal, which by Cor. 1 is impossible.

(20) *Cor.* 3. Hence, there is another point situated, in respect to the curve, precisely like the focus F, and may, therefore, be called another focus. Thus, if $CV = CF$, the point V is the other focus.

(21) *Cor.* 4. If different ellipses have the same transverse axis, the corresponding ordinates are proportional to each other.

That is, $FH : FH' :: GS : GS'$.

(Fig. 3.)

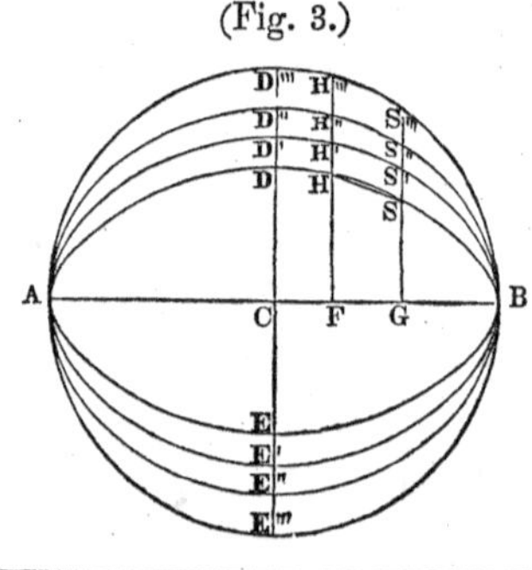

[a] Leg. 4. 10. Euc. 2. 5, Cor. [b] Leg. 4. 11. Euc. 1. 47.

For (17)	$FH^2 : GS^2 :: AF.FB : AG.GB.$
Also,	$FH'^2 : GS'^2 :: AF.FB : AG.GB.$
Therefore	$FH^2 : FH'^2 :: GS^2 : GS'^2.$
And[a]	$FH : FH' :: GS : GS'.$

(22) *Cor.* 5. It follows from the last of the foregoing proportions,[b] that $HH' : SS' :: FH : GS$.

(23) *Cor.* 6. Since OC is midway between AL and BI, it equals half their sum. But $BI+AL=BF+AF=AB$. Therefore OC, or its equal FD, is equal to AC, the semi-transverse axis.

(24) Prop. II. Theorem.

The square of an ordinate to the transverse axis, is to the rectangle of the corresponding abscissas, as the square of the conjugate axis is to the square of the transverse axis.

That is (Fig. 2), $RV^2 : AV.VB :: DE^2 : AB^2$.

For by similar triangles (IGF and ILM) we have

$$GF : LM=2AF :: IG : IL :: FB : AB.$$

And by similar triangles (LGF and LIK) we have

$$GF : IK=2FB :: LG : IL :: AF : AB.$$

Multiplying the proportions together,

$$GF^2 : 4AF.FB=(7)\ DE^2 :: AF.FB : AB^2.$$

But (17) $GF^2 : RV^2 :: AF.FB : AV.VB.$

Therefore, by equality of antecedents,[c]

$$RV^2 : AV.VB :: DE^2 : AB^2.$$

(24*a*) *Cor.* 1. The line GH is the parameter of the transverse axis.

For, as above, $GF^2 : DE^2 :: AF.FB=\frac{1}{4}DE^2 : AB^2.$

[a] Leg. 2. 12, Cor.
[b] Leg. 2. 6. Euc. 5, D. and 16.
[c] Leg. 2. 4. Euc. 5. 24.

Therefore, extracting roots,[a]

$$AB : \tfrac{1}{2}DE :: DE : GF = \tfrac{1}{2}GH.$$

Or, doubling the second and fourth terms,

$AB : DE :: DE : GH =$ the parameter, which we shall hereafter designate by the letter p.

(25) *Cor.* 2. If a circle be described on the transverse axis of an ellipse, an ordinate to the ellipse is to the corresponding ordinate to the circle, as the conjugate axis is to the transverse.

For (Fig. 3)

$$GS^2 : AG.GB = {}^{b}GS'''^2 :: DE^2 : AB^2.$$

Hence, extracting roots,

$$GS : GS''' :: DE : AB.$$

(26) *Cor.* 3. If the conjugate axis of an ellipse is equal to the transverse, the ellipse becomes a circle. For then the square of the ordinate becomes equal to the rectangle of the corresponding abscissas, which is a known property of the circle.[b]

(27) *Cor.* 4. The extremities of the conjugate axis of an ellipse lie in the curve. For

$$CD^2 = FD^2 - FC^2 = (23)\ AC^2 - FC^2 = {}^{c}AF.FB,$$

which, by (7), is equal to the square of the semi-conjugate axis.

(28) Prop. III. Theorem.

The sum of two lines drawn from the foci of an ellipse to any point in the curve, is equal to the transverse axis.

That is, $VM + FM = AB$.

Make the arc AN equal to BM, join FN, draw the ordinates NR and MS, the semi-conjugate axis DC, and the focal tangent TI, and produce NR, MS, and CD to G, P, and O.

---

[a] Leg. 2. 12, Cor. [b] Leg. 4. 23, Cor. Euc. 6. 8, Cor.
[c] Leg. 4. 10. Euc. 2. 5, Cor.

(Fig. 4.)

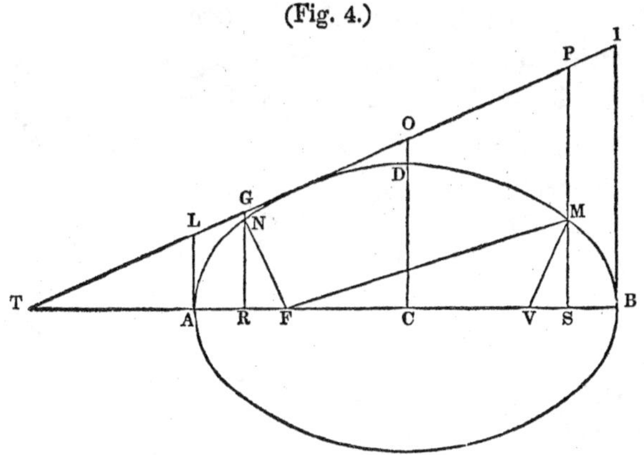

Then, since CO is midway between GR and PS, it is equal to half their sum. Therefore

$$GR+PS=2OC=(23)\ AB.$$

But (2) $GR+PS=FN+FM=(19)\ VM+FM.$

Therefore $VM+FM=AB.$

(29) *Scholium.* The property proved in this proposition furnishes the definition of the ellipse in many treatises. It also affords a ready method of describing the curve mechanically. Take a thread of the length of the transverse axis, and fasten one of the ends to each of the foci. Then carry a pencil round by the thread, keeping it always stretched, and its point will describe the ellipse. For in every position of the pencil, the sum of the distances to the foci will be equal to the entire length of the string.

(30) Prop. IV. Theorem.

Two lines drawn from the foci to any point in the curve, make equal angles with a tangent to the curve at that point.

That is, FM and VM make equal angles with the tangent TU, or FMT=VMU.

(Fig. 5.)

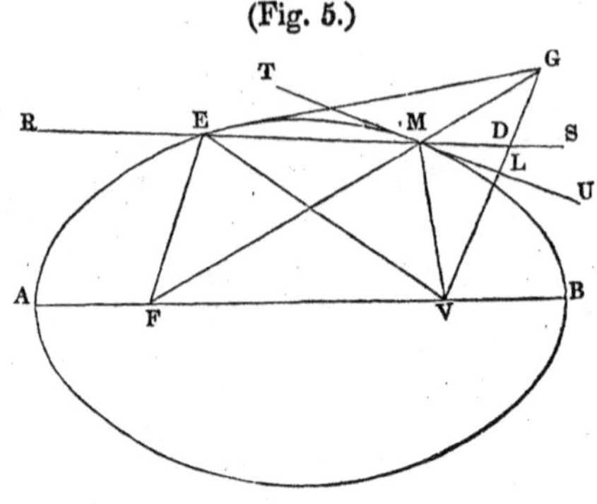

If not let them make equal angles with RS, so that FMR=VMS.

Since there cannot be two different tangents to a curve at the same point, RS must cut it. Let it cut in M and E.

Produce FM to G, making MG=MV, and join GV, GE, EF, and EV.

The angle GMS=RMF= (by supposition) VMS. Then, in the triangles GMD and VMD, GM=MV, and MD is common, and the angle GMD=VMD; therefore GD=VD, and the angle GDM =VDM. Hence, in the triangles EGD and EVD, the sides GD and ED=VD and ED, and the angle GDE=VDE; therefore GE=EV, and EF+GE=EF+EV=(28) FM+MV=FG; that is, two sides of a triangle are equal to the third side, which is impossible.[a] In the same manner it may be shown, that no other line that TU makes equal angles with FM and MV, and consequently TU does.

(31) *Cor.* 1. Hence, to draw a tangent to the curve at any point M, join MF and MV, and bisect the exterior angle VMG.

(32) *Cor.* 2. GV is perpendicular to TU, and is bisected at the point L.

(33) *Cor.* 3. FG=AB, since each is equal to FM+MV.

---

[a] Leg. 1. 7. Euc. 1. 20.

(34) Prop. V. Theorem.

If a line be drawn from either focus perpendicular to a tangent to the curve at any point, the distance of its intersection from the centre is equal to the semi-transverse axis.

That is, CL=AC.

(Fig. 6.)

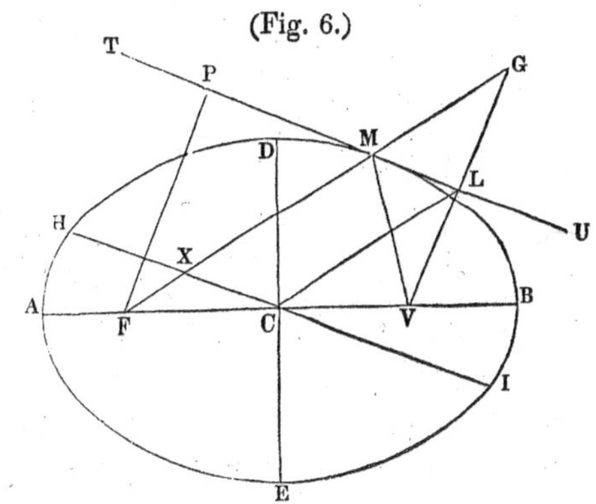

Since FV is bisected in C, and (32) GV in L, CL is parallel[a] to FG, and the triangles LCV and GFV are similar. Hence

CV : FV :: CL : GF.

But CV=$\frac{1}{2}$FV,

Therefore CL=$\frac{1}{2}$GF=(33) $\frac{1}{2}$AB=AC.

(35) *Cor.* 1. Hence, a circle described on the transverse axis with the centre C, will pass through the intersections L and P; and conversely, if from any point in the circumference of such a circle, two lines be drawn at right angles to one another; and if one of them pass through one of the foci, the other will be tangent to the curve.

(36) *Cor.* 2. Hence, a diameter HI, parallel to TU, would cut off a part MX of the line MF, equal to the semi-transverse axis.

For[b] MX=CL=AC.

(37) *Cor.* 3. Since CL is parallel to FM, the angle CLM=FMP=(30) VML.

---

[a] Leg. 4. 16. Euc. 6. 2. [b] Leg. 1. 28. Euc. 1. 34.

(38) Prop. VI. Theorem.

The rectangle of the two perpendiculars drawn from the foci to any tangent to the curve, is equal to the square of the semi-conjugate axis.

That is, $VL.FP = CD^2$.

(Fig. 7.)

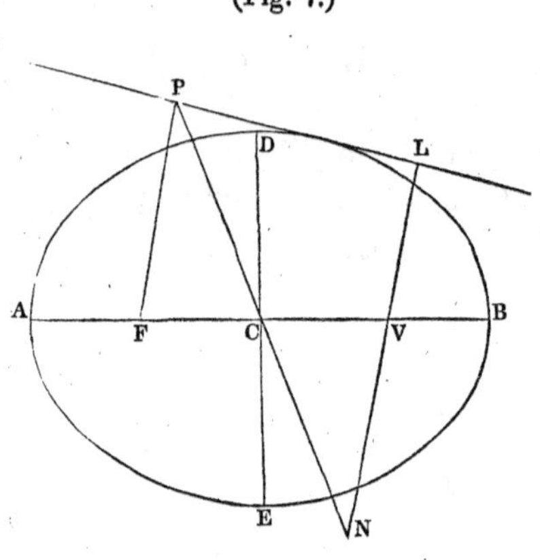

Join PC, and produce PC and LV till they meet in N.

Then will the triangles PFC and NVC be similar and equal, since the angle PCF=NCV, and FPC=[a]the alternate angle VNC, and the side FC=VC. Therefore CN=CP, and (35) the point N is in the circumference of a circle described on AB as a diameter. Consequently,[b] NV.VL=AV.VB. Or, since NV=PF, $PF.LV = AV.VB = (7)\ CD^2$.

(38*a*) Prop. VII. Theorem.

If at any point in the curve a tangent and ordinate be drawn, meeting either axis produced, half that axis is the mean proportional between the distances of the two intersections from the centre.

That is, CS : CA :: CA : CT.

Or, CG : CD :: CD : CH.

---

[a] Leg. 1. 20, Cor. 2. Euc. 1. 29. [b] Leg. 4. 28, Cor. Euc. 3. 35.

(Fig. 8.)

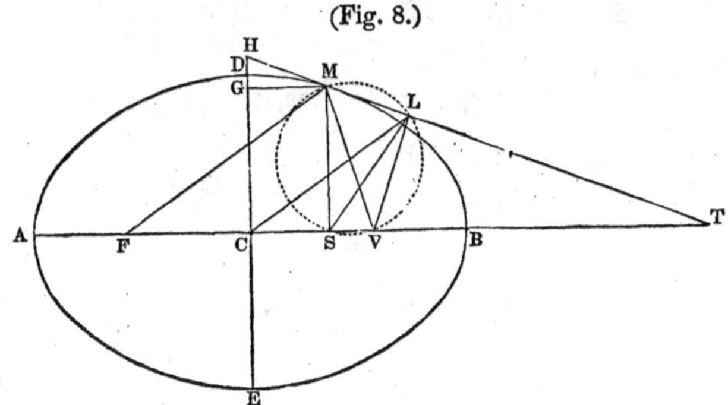

Connect M with the foci F and V, draw VL perpendicular to the tangent, and join CL and SL.

Since MSV and MLV are both right angles, each is an angle in a semicircle,[a] and consequently, a circle described on MV as a diameter, would pass through L and S. Then must the angles VML and VSL be equal, being in the same segment,[b] or measured by the same arc.

But (37) VML = CLM. Therefore the angles VSL and CLM are equal, as also their supplements CSL and CLT. Hence, the triangles LCT and SCL are similar, for the angle CSL = CLT, and the angle at C is common.

Therefore CS : CL :: CL : CT.

But (34) CL=CA.

Therefore CS : CA :: CA : CT,

which proves the proposition in respect to the transverse axis.

(38*b*) Again, since when three numbers are in continued proportion, the first is to the third as the square of either antecedent is to the square of its consequent, we have from the last proportion

$CS^2 : CA^2 :: CS = GM : CT ::$ (sim. tri.) $GH : CH$.

Hence, by division,[c] $CA^2 - CS^2 : CA^2 :: CG : CH$.

---

[a] Leg. 3. 18, Cor. 2. Euc. 3. 31. [b] Leg. 3. 18. Euc. 3. 21.
[c] Leg. 2. 6. Euc. 5. 17.

But (24) $AS.SB =$ [a]$CA^2 - CS^2 : CA^2 :: MS^2 = CG^2 : CD^2$.

Therefore, by equality of ratios,

$$CG : CH :: CG^2 : CD^2;$$

which gives, $CG.CH = CD^2$.

Or, $CG : CD :: CD : CH$.

(39) PROP. VIII. THEOREM.

If different ellipses have the same transverse axis, the corresponding sub-tangents are equal to one another.

(Fig. 9.)

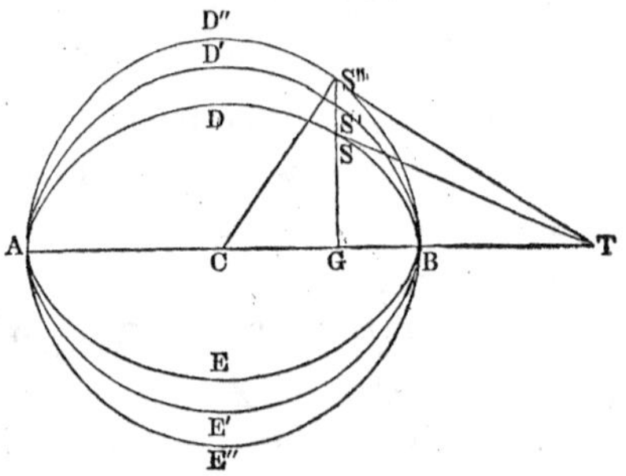

Let EBDA, E′BD′A, &c. be any number of ellipses described on AB as the transverse axis, SG produced any ordinate to each, and ST, S′T, &c. tangents at the points where the ordinate meets the curves.

Then for each of them we have (38*a*)

$$CG : CA :: CA : CT,$$

and since the first three terms are the same for all, the fourth must be likewise, which, diminished by CG, gives the sub-tangent TG.

(39*a*) *Cor.* 1. Hence, we may draw a tangent at a given point in the curve, without knowing the foci.

Let S be the given point. On AB describe a circle; draw the ordinate SG, and produce it till it meets the circle in S″. Draw[b] S″T tangent to the circle at S″, and join TS.

(39*b*) *Cor.* 2. $CG.GT = AG.GB$, each being equal[c] to $S''G^2$.

---

[a] Leg. 4. 10. Euc. 2. 5, Cor. [b] Leg. 3, Prob. 14. Euc. 3. 17.
[c] Leg. 4. 23. Euc. 6. 8, Cor. and 13.

(40) Prop. IX. Theorem.

If ordinates be drawn from the extremities of any two conjugate diameters, the sum of the squares of the parts of the transverse axis intercepted between the ordinates and the centre, is equal to the square of the semi-transverse axis; and the sum of the squares of the ordinates is equal to the square of the semi-conjugate axis.

That is, $CR^2+CS^2=AC^2$, and $PS^2+UR^2=CE^2$.

Draw the tangents PT and UV, meeting the transverse axis in T and V. Then

$CS.CT=CR.CV$,

each being equal (38*a*) to AC^2 or BC^2;

(Fig. 10.)

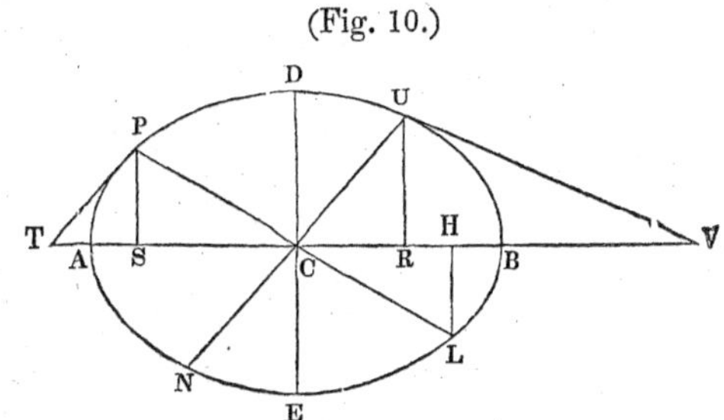

Therefore[a] $CS : CR :: CV : CT$.

But, since (9) UV is parallel to PC, and UC to PT, the triangles PTC and UCV are similar; and, hence, similarly divided by the ordinates PS and UR.

Therefore $CR : TS :: CV : CT$.

Then, by equality of ratios, we have

$CS : CR :: CR : TS$, and $CS.TS=CR^2$.

But (39*b*) $CS.TS=AS.BS=$[b]AC^2-CS^2.

Therefore $AC^2-CS^2=CR^2$; or, $CR^2+CS^2=AC^2$.

(41) Again, since $AS.BS=CR^2$, and, in like manner, $AR.BR=CS^2$, we have (24)

$CE^2 : AC^2 :: PS^2 : CR^2$.

---

[a] Leg. 2. 2. Euc. 6. 16. [b] Leg. 4. 10. Euc. 2. 5, Cor.

And $CE^2 : AC^2 :: UR^2 : CS^2$.

Therefore $CE^2 : AC^2 :: PS^2+UR^2 : CR^2+CS^2$.

But (40) $CR^2+CS^2=AC^2$.

Therefore $PS^2+UR^2=CE^2$. (See Appendix, F.)

(42) Prop. X. Theorem.

The sum of the squares of any pair of conjugate diameters is equal to the sum of the squares of the two axes.

That is, $UN^2+PL^2=AB^2+DE^2$.

(Fig. 11.)

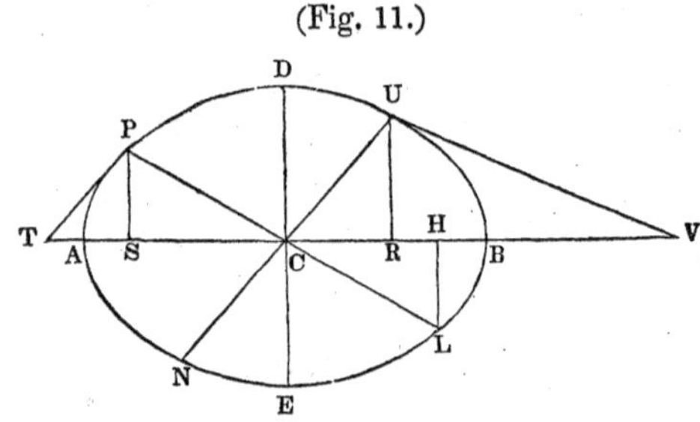

For[a] $UC^2+PC^2=CR^2+CS^2+UR^2+PS^2$.

But (40) $CR^2+CS^2=AC^2$, and $PS^2+UR^2=CE^2$.

Therefore $UC^2+PC^2=AC^2+CE^2$.

Or, $UN^2+PL^2=AB^2+DE^2$.

(43) Prop. XI. Theorem.

Any parallelogram circumscribed about an ellipse, having its sides parallel to two conjugate diameters, is equal to the rectangle of the two axes.

That is, $GHIK=LMNO=AB\times DE$.

[a] Leg. 4. 11. Euc. 1. 47.

(Fig. 12.)

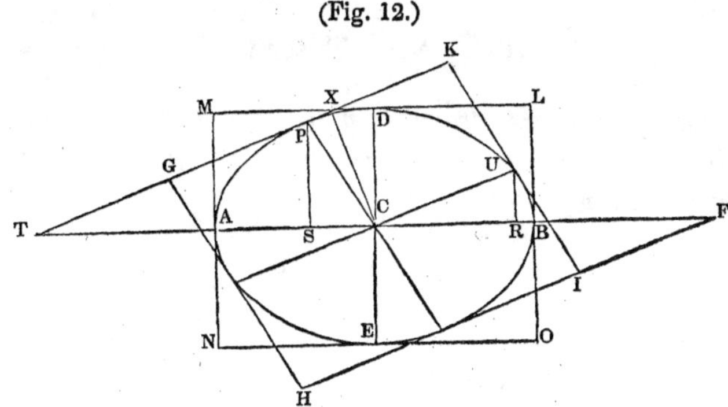

Draw CX at right angles to TK; then will the triangles TCX and CUR be similar.

Now (24) $DC^2 : AC^2 :: UR^2 : AR.RB =$
(as shown in 40 and 41) CS^2.

Therefore, extracting roots,

$DC : AC :: UR : CS.$

Or, alternately, $AC : CS :: DC : UR.$

But (38*a*) $AC : CS :: CT : AC.$

Therefore $DC : UR :: CT : AC,$

or[a] $UR.CT = DC.AC = \frac{1}{4}AB.DE.$

Also $CU : UR :: CT : CX,$

or $UR.CT = CU.CX =$[b] $\frac{1}{4}GHIK.$

Therefore $GHIK = AB.DE.$

(44) *Cor.* 1. $AC : CX :: CU : DC,$ or $CU.CX = AC.DC.$

---

[a] Leg. 2. 2. Euc. 6. 16. [b] Leg. 4. 5.

(45) PROP. XII. THEOREM.

The rectangle of two lines, drawn from the foci of an ellipse to any point in the curve, is equal to the square of half the diameter parallel to the tangent at that point.

That is, $FM.VM=CH^2$.

Draw the tangent LMP, and the perpendiculars to it FL, MN, and VP. Then will the triangles PMV, SMN, and FML be similar (30).

(Fig. 13.)

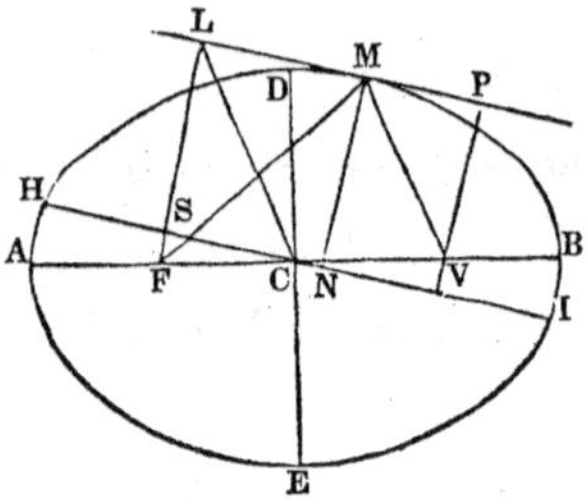

Therefore

$MS : MN :: FM : FL.$

And also,

$MS : MN :: VM : VP.$

Multiplying the proportions together,

$MS^2 : MN^2 :: FM.VM : FL.VP.$

But (36) $MS^2=AC^2$, and (38) $FL.VP=CE^2$.

Therefore $AC^2 : MN^2 :: FM.VM : CE^2$.

Now (43) $MN.CH=AC.CE$.

Therefore $AC : MN :: CH : CE$.

Or, squaring, $AC^2 : MN^2 :: CH^2 : CE^2$.

Hence, by equality of ratios, $CH^2 : CE^2 :: FM.VM : CE^2$.

And, consequently, $FM.VM=CH^2$.

---

[a] Leg. 4. 5.

(46) Prop. XIII. Theorem.

If at one of the vertices of an ellipse a tangent be drawn meeting any diameter produced, and also from the same point an ordinate to that diameter, the semi-diameter is the mean proportional between the distances of the two intersections from the centre.

That is, CE : CI :: CI : CP.

(Fig. 14.)

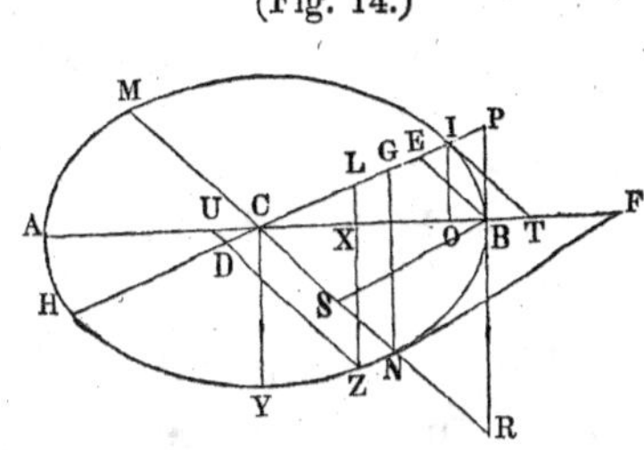

Draw IT and IO tangent and ordinate to the curve at I. Then, by similar triangles,

CE : CI :: CB : CT.

But (38*a*) CO : CB :: CB : CT.

And, by similar triangles,

CO : CB :: CI : CP.

Hence, by equality of ratios,

CE : CI :: CI : CP.

(47) *Cor.* 1. Hence, $CI^2 : CP^2$ :: CE : CP; and, in like manner, $CN^2 : CR^2$ :: CS : CR.

(48) *Cor.* 2. The lines CP and CT are similarly divided, the former in the points I and E, and the latter in B and O; and, consequently, [a]lines joining EO, IB, and PT, would be parallel.

[a] Leg. 4. 16. Euc. 6. 2.

(49) *Cor.* 3. The area of the triangle CIT=CBP, and CEB=COI; for the angle at C is common, and the sides about it reciprocally proportional.[a] In like manner, CBR=CNF.

(50) *Cor.* 4. The area of the triangle CNG=CBP or CIT.*

For[b] CNG : CRP :: $CN^2 : CR^2$ :: (Cor. 1) CS=BE : CR :: (sim. tri.) BP : PR :: [c]CBP : CRP.

Hence, since CNG and CBP have the same ratio to CRP, they are equal to one another.

(51) *Cor.* 5. IOBP=IOT.

For (49) CIT=CBP, and taking CIO from each, the remainders must be equal.

(52) *Cor.* 6. The triangle UZX=PBXL, (Z being any point in the curve.)

For CB : BP :: CX : XL :: [d]CB+CX=AX : BP+XL.

Also, CB : BP :: CO : OI :: CB+CO=AO : BP+OI.

Therefore AX : AO :: BP+XL : BP+OI.

Or,[e] AX.XB : AO.OB :: $\overline{BP+XL}$.XB : $\overline{BP+OI}$.OB.

But

$\overline{BP+XL}$.XB=[f]2PLXB, and $\overline{BP+OI}$.OB=2PIOB=(51) 2IOT.

Therefore

PLXB : IOT :: AX.XB : AO.OB :: (17) $ZX^2 : OI^2$ :: [d]UZX : IOT.

Hence, PLXB=UZX, since both have the same ratio to IOT.

(53) *Cor.* 7. DZL=ICT−DUC.

For (49)

ICT=CBP=CLX+PLXB=(52) CLX+UZX=DUC+DZL,

since the part ZDCX is common to both UZX and DZL.

Hence, DZL=ICT−DUC.

---

[a] Leg. 4. 24, Cor. Euc. 6. 15. [b] Leg. 4. 25. Euc. 6. 19.
[c] Leg. 4. 6, Cor. Euc. 6. 1. [d] Leg. 2. 10. Euc. 5. 12.
[e] Leg. 2. 8. Euc. 5. 15 and 16. [f] Leg. 4. 7.

* HI and MN being conjugate diameters.

(54) Prop. XIV. Theorem.

The square of any diameter is to the square of its conjugate, as the rectangle of its abscissas is to the square of the corresponding ordinate.

That is, $HI^2 : MN^2 :: HD.DI : DZ^2$.

(Fig. 15.)

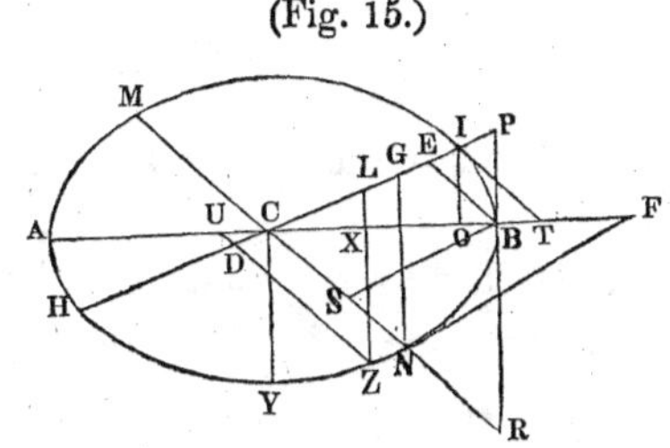

For[a] $CN^2 : DZ^2 :: CNG = (50)\ ICT : DZL = (53)\ ICT - DUC$.

But[a] $ICT : DUC :: CI^2 : CD^2$.

Hence,[b] $ICT : ICT - DUC :: CI^2 : CI^2 - CD^2 =$ [c]$HD.DI$.

Therefore, by equality of ratios,

$$CN^2 : DZ^2 :: CI^2 : HD.DI.$$

Or, alternately, $CI^2 : CN^2 :: HD.DI : DZ^2$.

Or, $HI^2 : MN^2 :: HD.DI : DZ^2$.

(55) *Cor.* Hence, all chords parallel to any diameter are bisected by its conjugate; and, conversely, a line bisecting two or more parallel chords is a diameter.

(56) Prop. XV. Problem.

To draw a tangent to an ellipse from a given point without the curve.

---

[a] Leg. 4. 25. Euc. 6. 19. [b] Leg. 2. 6. Euc. 5, D.
[c] Leg. 4. 10. Euc. 2. 5, Cor.

(Fig. 16.)

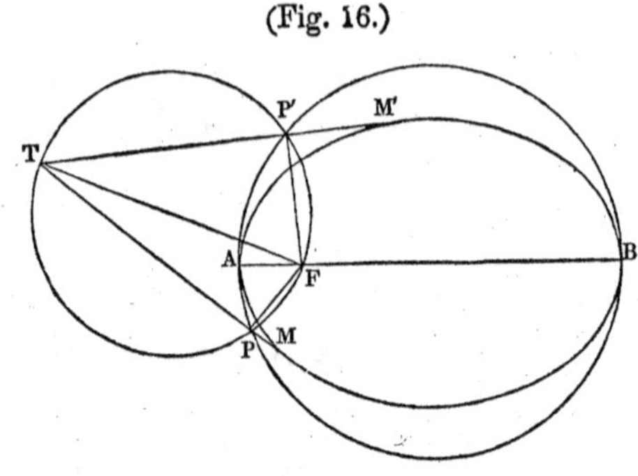

Let AMB be the given ellipse, AB the transverse axis, F one of the foci, and T the given point.

Join TF, and upon it and AB as diameters, describe the circles TPF and APB, cutting each other in P and P′. The lines TPM and TP′M′ drawn through the points of intersection will be tangents to the ellipse.

Join FP. The angle FPT is[a] a right angle, and FP perpendicular to TM. Now, since from the point P, in the circumference of the circle described on the transverse axis, there are drawn two lines, PF and PM, at right angles to one another, and one of them (PF) passes through the focus, the other must (35) be tangent to the ellipse.

(57) Prop. XVI. Problem.

To find the centre, axes, and foci of a given ellipse.

(Fig. 17.)

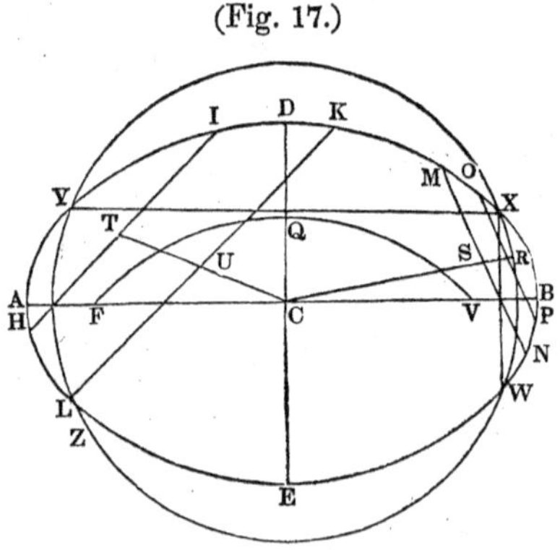

Let BDAE be the given ellipse.

Draw any two pairs of parallel chords HI and LK, MN and OP; bisect them in T, U, R, and S; join TU and RS, and produce the lines till they meet in C. Both these lines being (55) diameters, the point C must be the centre.

[a] Leg. 3. 18, Cor. 2. Euc. 3. 31.

From the centre C, and with any convenient radius, describe a circle cutting the ellipse in any points W, X, Y, and Z. Draw WX and XY, and at right angles to them respectively, draw AB and DE through the centre C. Since these lines pass through the centre, they are diameters; and since they bisect [a]WX and XY at right angles, they divide the ellipse into two similar parts, and therefore are (19) axes.

From E, the extremity of the conjugate axis as a centre, and with a radius equal to the semi-transverse axis, describe the arc FQV, cutting the transverse axis in F and V. These points are (23) the foci.

(57*a*) Prop. XVII. Theorem.

An ellipse may be formed by the mutual intersection of a cone and a plane.

Suppose the ellipse in Prop. I., with no change of letters, to be placed upon the cone CDLE in the manner of a collar, with its plane perpendicular to the triangular section CDE, the latter being perpendicular to the base of the cone, and passing through A, B, and C. Now, suppose the point A to slide up or down on the line CD, and B on CE, till the point G shall lie in the surface of the cone; a condition which is evidently possible, whatever be the nature of the curve AGBH. We assert that then will any other point R in the ellipse also lie in the surface of the cone.

(Fig. 18.)

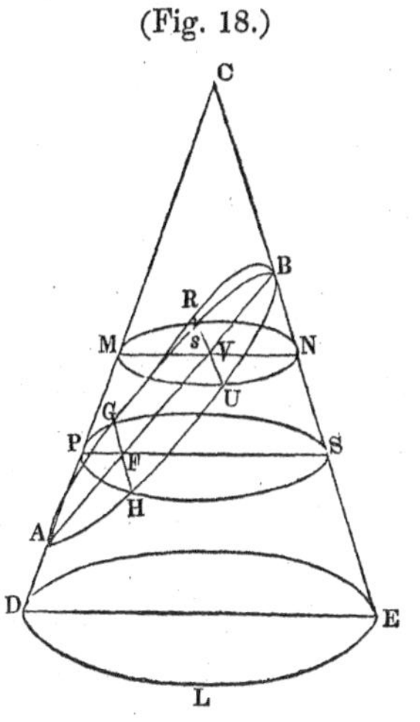

If not, it must lie either within or without the cone. Let it be supposed to lie without, and that the ordinate

---

[a] Leg. 3. 6. Euc. 3. 3.

RV cuts the surface of the cone at s Through G and s let the circular sections PGSH and MsNU be made to pass, parallel to the base, and cutting the triangular section in PS and MN. The lines GF and RV being perpendicular to the plane CDE, must also be perpendicular to PS and MN.

By sim. tri. AFP and AVM, PF : MV :: AF : AV.

And by sim. tri. BFS and BVN, FS : VN :: FB : VB.

Multiplying the proportions together,

PF.FS : MV.VN :: AF.FB : AV.VB.

But[a] PF.FS$=$GF^2, and MV.VN$=s$V^2.

Therefore GF^2 : sV^2 :: AF.FB : AV.VB.

But (17) GF^2 : RV^2 :: AF.FB : AV.VB.

Therefore sV^2=RV^2, and sV=RV, which is impossible.

In the same manner it may be shown, that the point R cannot lie within the cone, and, consequently, it lies in the surface. And since R is any point in the ellipse, the whole curve must lie in the surface of the cone.

---

[a] Leg. 4. 23, Cor. Euc. 6. 13.

CHAPTER III.

OF THE PARABOLA.

---

(58) Prop. I. Theorem.

The squares of ordinates to the transverse axis of a parabola are to each other as the corresponding abscissas.

That is, $GF^2 : RV^2 :: AF : AV$.

(Fig. 19.)

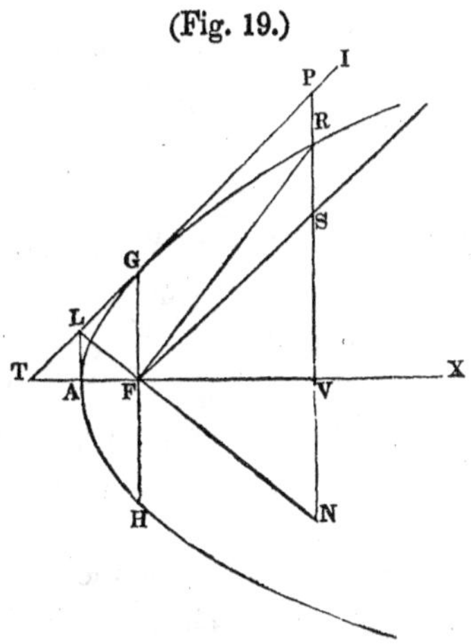

Let F be the focus

Draw the focal tangent IT, and FS parallel to it; from the vertex A draw AL at right angles to AX; join LF, and produce RV both ways till it meets IT and LF produced in P and N.

Then, because (1) FG=FT, and (15[a]) AL=AF, we have also by similar triangles, as in Prop. I. of the ellipse, AL=AF, SV=FV, and VN=FV.

By similar triangles, LGF and LPN, we have

$$GF : PN :: LG : LP :: {}^{a}AF : AV.$$

Multiplying the first couplet by GF=PS,

$$GF^2 : PS \times PN :: AF : AV.$$

But, as in Prop. I. of the ellipse, $PS \times PN = RV^2$.

Therefore, $GF^2 : RV^2 :: AF : AV$.

---

[a] Leg. 4. 15, Cor. 2. Euc. 6. 2.

Schol. If we suppose the parabola to have anotner vertex at an infinite distance in the direction AX, this proposition will be the same as Prop. I. of the ellipse.

(59) *Cor.* 1. The line GH is the parameter of the transverse axis.

For since (2) in the parabola the angle GTF=45°, AT=AL =AF. Hence FT=2AF, FG=2AF, and GH=4AF.

Substituting 2AF in place of GF in the last proportion, we have

$$4AF^2 : RV^2 :: AF : AV.$$

Dividing the first and third terms by AF, we obtain

$$4AF : RV^2 :: 1 : AV.$$

$$\text{Or, } AV : 1 :: RV^2 : 4AF=GH.$$

Or, by transferring the factor RV,

$$AV : RV :: RV : GH.$$

(60) *Cor.* 2. By the third proportion in the foregoing corollary we learn that $RV^2=AV.4AF=AV.GH$; that is, the square of any ordinate is equal to the abscissa multiplied by four times the focal distance, or by the parameter.

(60*a*) *Cor.* 3. The two portions of the curve lying on the opposite sides of the transverse axis are symmetrical.

Props. II. and III. of the ellipse are applicable to the parabola only upon the supposition that its transverse axis is infinite, and that it has another focus infinitely distant.

(61) Prop. IV. Theorem.

Two lines drawn from any point in the curve, one to the focus and the other parallel to the transverse axis, make equal angles with a tangent to the curve at that point.

That is, FM and VM make equal angles with TU, or FMT=VMU.

If not, let them make equal angles with RS, so that RMF= VMS.

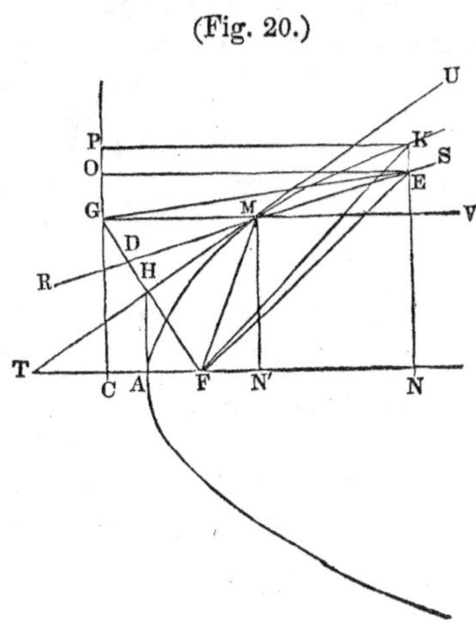

(Fig. 20.)

Since there cannot be two different tangents to a curve at the same point, RS must cut it and fall within, as at some point E. Through E draw the ordinate KN cutting the curve in K. Produce VM till it meets the directrix in G; join GF, GE, EF, and KF, and draw KP and EO parallel to AN.

The angle GMR = VMS = (by supposition) RMF. Then, in the triangles GMD and FMD, GM=(1) FM and MD is common, and the angle GMD= FMD; therefore GD=FD, and the angle GDM=FDM. Hence, in the triangles EGD and EFD, the two sides GD and ED are equal to FD and ED, and the angle GDE=FDE; therefore EG= EF. Now KF, being opposite to the greater angle of the triangle EKF, is greater [a]than EF, and is therefore greater than EG. But KF=(1) KP=EO; therefore EO is greater than EG; that is, one of the perpendicular sides of a right-angled triangle is greater than the hypotenuse, which is impossible.

In the same manner it may be shown that no other line but TU makes equal angles with FM and VM, and consequently TU does.

(62) *Cor.* 1. Hence, to draw a tangent to the curve at any point M, join MF, draw MG parallel to the transverse axis, and bisect the angle FMG.

(63) *Cor.* 2. GF is perpendicular to MT, and they mutually bisect each other at the point H, as is evident from the equal triangles GHM and FHT.

---

[a] Leg. 1. 13. Euc. 1. 19.

(64) *Cor.* 3. FM=FT.

(65) *Cor.* 4. AN′=AT, or TN′=2AN′; that is, every subtangent is bisected at the vertex, or is equal to twice the abscissa.

For CN′=GM=FM=FT.

And (1) CA =AF.

Therefore, subtracting equals,

AN′=AT, or 2AN′=TN′.

(66) *Cor.* 5. If different parabolas have the same transverse axis, the corresponding subtangents will be equal to one another. For in each case the sub-tangent will be equal to twice the abscissa.

(67) *Cor.* 6. Hence we may draw a tangent at a given point, as M, in the curve, without knowing the focus (62); viz., draw the ordinate MN′, make AT equal to AN′, and join TM.

(68) Prop. V. Theorem.

If a line be drawn from the focus perpendicular to a tangent to the curve at any point, a tangent at the vertex will pass through the point of intersection.

That is, if FH is perpendicular to TM, a tangent to the curve at A will pass through H.

(Fig. 21.)

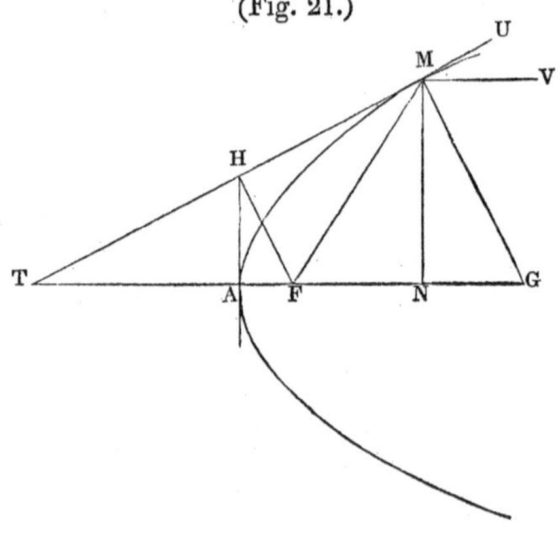

Since (63) TM is bisected in H and (65) TN in A, AH is parallel to MN.

Hence it is perpendicular to AN, and consequently tangent to the curve at A. And as there cannot be two tangents to a curve at the same point, a tangent at A must pass through H

(68*a*) *Cor.* 1. Hence, if a line be drawn from F to any point H in the tangent at the vertex, a line HM drawn from H perpendicular to FH will touch the curve.

(68*b*) *Cor.* 2. The sub-normal NG=2AF, and is therefore a constant quantity, wherever in the curve the point M be taken.

For, by similar triangles,

$$TH : TM :: AH : MN :: AF : NG.$$

But (63) TM=2TH, therefore NG=2AF.

(68*c*) *Cor.* 3. $FH^2=AF.FM$.

For[a] AF : FH :: FH : FT=FM.

(68*d*) *Cor.* 4. $TM^2=AN.4FM$, or $\frac{TM^2}{AN}=4FM$.

For, by similar triangles, TF : TH :: TH : TA=AN.

Therefore, $TH^2=AN.TF=AN.FM$.

And, consequently, $TM^2=AN.4FM$.

Prop. V. of the ellipse is true also of the parabola, if we regard its transverse axis as infinite. For the circumference of a circle whose diameter is infinite is a straight line, so that AH of the preceding figure may be considered as an arc of a circle described on the transverse axis.

Now it has been shown (68) that the perpendicular FH meets TM in this circumference, and therefore at a distance from the centre equal to the radius, that is, to half the transverse axis.

Props. VI., VII., IX., X., XI., and XII. of the ellipse demonstrate properties to which there are none analogous in the parabola.

Prop. VIII. of the ellipse corresponds to (66).

---

[a] Leg. 1. 13. Euc. 1. 19.

(69) Prop. XIII. Theorem.

If at the vertex of a parabola a tangent be drawn meeting any diameter produced, and also from the same point an ordinate to that diameter, the distances of the intersections from the curve measured on the diameter will be equal.[a]

That is, MP=ME.

(Fig. 22.)

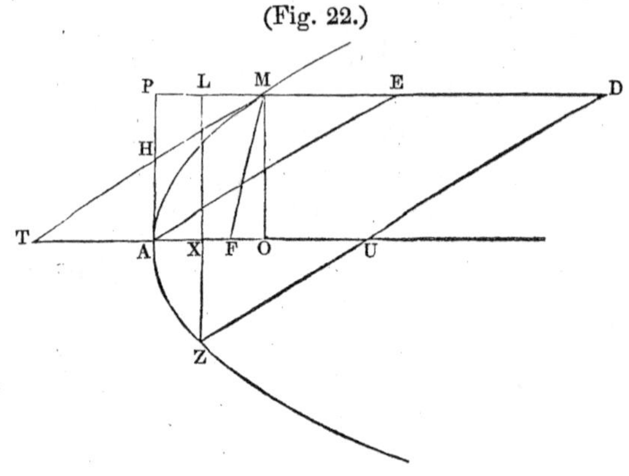

For, since the angle PHM=AHT, and HPM=HAT, and (63 and 68) the side TH=HM, we have AT=MP.

But[b] AT=ME.

Therefore MP=ME.

(70) *Cor.* 1. The triangle HAT=HPM.

(71) *Cor.* 2. The lines EP and OT are similarly divided, so that lines joining EO, MA, and PT, would be parallel.

(72) *Cor.* 3. The triangle OMT=APE=[c]PAOM=[d]MTAE.

---

[a] A line drawn parallel to the transverse axis from any point in the curve is called a diameter.

[b] Leg. 1. 28. Euc. 1. 34.

[c] Leg. 4. 2. Euc. 1. 41.

[d] Leg. 4. 1. Euc. 1. 36.

(73) *Cor.* 4. The triangle UZX=PAXL, (Z being any point in the curve.)

For (58) $ZX^2 : OM^2 :: AX : AO ::$ [a]PAXL : PAOM.

But (sim. tri. UZX and OMT) we have

[b]$ZX^2 : OM^2 :: UZX : OMT.$

Therefore, PAXL : PAOM :: UZX : OMT.

But (72) OMT=PAOM; therefore UZX=PAXL.

(74) *Cor.* 5. DZL=DUAP=DUTM.

For DZL=DUXL+UZX=(73) DUXL+PAXL=DUAP.

(75) Prop. XIV. Theorem.

The squares of ordinates to any diameter of a parabola are to each other as the corresponding abscissas.

That is, $EA^2 : DZ^2 :: ME : MD.$

(Fig. 23.)

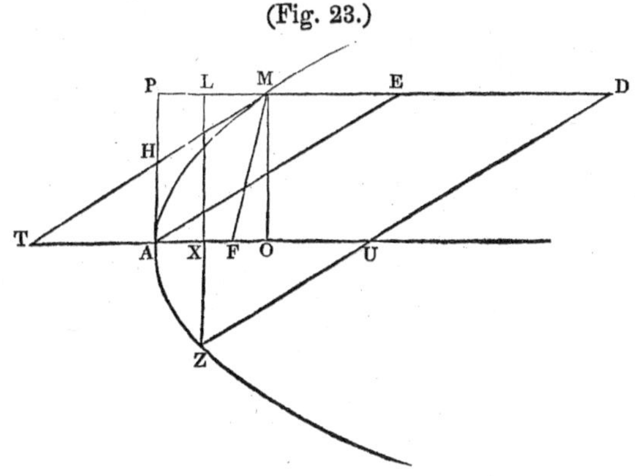

For [c]$EA^2 : DZ^2 ::$ APE=(72) MTAE : DZL= (74) DUTM :: [a]ME : MD.

---

[a] Leg. 4. 3. Euc. 6. 1. [b] Leg. 4. 25. Euc. 6. 19.
[c] Leg. 4. 25. Euc. 6. 19.

(76) *Cor.* 1. Hence all chords parallel to a tangent at any point of a parabola are bisected by a diameter terminating at that point; and conversely, a line bisecting two or more parallel chords is a diameter.

(76*a*) *Cor.* 2. ME : AE :: AE : 4FM.

For (68*d*) $TM^2 = AO.4FM$.

But $TM = AE$, and $AO = ME$.

Hence $AE^2 = ME.4FM$.

(77) Prop. XV. Problem.

To draw a tangent to a parabola from a given point without the curve.

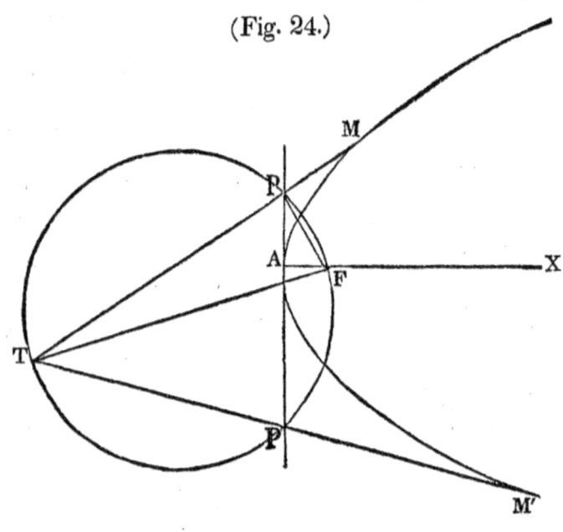

(Fig. 24.)

Let M'AM be the given parabola, AX the transverse axis, F the focus, A the vertex, and T the given point. Join TF, and upon it as a diameter describe the circle TPF, cutting PAP', the tangent at the vertex, in the points P and P'. The lines TPM and TP'M' drawn through the points of intersection will be tangents to the curve.

Join FP. The angle TPF is[a] a right angle, and FP perpendicular to PM. Then, since from a point P in the tangent AP, a line PM is drawn perpendicular to FP, it is (68*a*) a tangent to the curve.

[a] Leg. 3. 18, Cor. 2. Euc. 3. 31.

(78) Prop. XVI. Problem.

To find the axis and focus of a parabola.

(Fig. 25.)

Let GEAD be the given parabola.

Draw any two parallel chords DE and BC, bisect them in L and N, and through the points of bisection draw the line MU. Then (76) will MU be a diameter. Draw DG at right angles to MU, bisect it in P, and through P draw AX at right angles to DG. The line AX is the transverse axis (60*a*), since it bisects the chord DG at right angles.

Through M draw MH parallel to DE, through A draw AH perpendicular to AX, and from H, the point of their intersection, draw HF perpendicular to HM. Then, since AH a tangent at the vertex, MH a tangent at the point M, and HF a perpendicular to the latter, intersect each other in the same point H, the point F must (68*a*) be the focus.

(78*a*) Prop. XVII. Theorem.

A parabola may be formed by the mutual intersection of a cone and a plane.

(Fig. 26.)

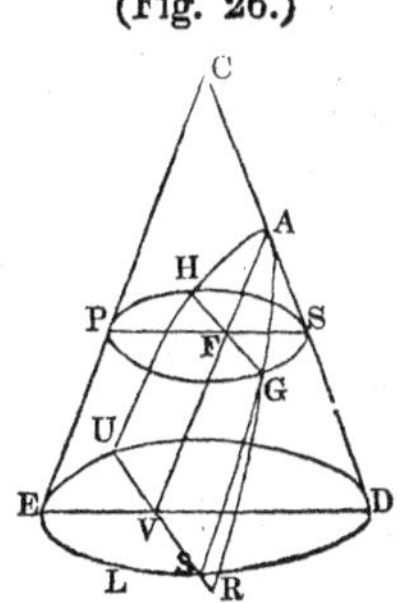

Suppose the parabola in Prop. I., with no change of letters, to be placed upon the cone CDLE in the manner of a collar, with its plane perpendicular to the triangular section CDE, and parallel to the side CE, the latter section being perpendicular to the base of the cone and passing through C and A.

Now, suppose the plane RGAH to move parallel to itself, yet so as to keep the point

A on the line CD till the point G shall lie in the surface of the cone, a condition which is evidently possible, whatever be the nature of the curve RGAH. We assert that then will any other point R in the parabola also lie in the surface of the cone.

If not, it must lie either within or without the cone. Let it be supposed to lie without, and that the ordinate RV cuts the surface of the cone at *s*. Through G and *s* let the circular sections PSGH and DUE*s* be made to pass, parallel to the base, and cutting the triangular section in PS and DE. The lines GF and RV being perpendicular to the plane CDE, must also be perpendicular to PS and ED.

By sim. tri., AFS and AVD,

$$AF : AV :: FS : VD.$$

Multiplying the last couplet by PF=[a]EV,

$$AF : AV :: PF.FS : EV.VD.$$

But[b] $PF.FS=FG^2$, and $EV.VD=sV^2$.

Therefore $AF : AV :: GF^2 : sV^2$.

But (58) $AF : AV :: GF^2 : RV^2$.

Therefore $sV^2=RV^2$, and $sV=RV$, which is impossible.

In the same manner it may be shown that the point R cannot lie within the cone, and consequently it lies in the surface. And since R is any point in the parabola, the whole curve must lie in the surface of the cone.

---

[a] Leg. 1. 28. Euc. 1. 34. [b] Leg. 4. 23, Cor. Euc. 6. 13.

CHAPTER IV.

OF THE HYPERBOLA.

---

(79) PROP. I. THEOREM.

The squares of ordinates to the transverse axis of an hyperbola are to each other as the rectangles of the corresponding abscissas.

That is, $GF^2 : RV^2 :: AF.FB : AV.VB$.
Or, $GF^2 : R'V'^2 :: AF.FB : AV'.V'B$.

(Fig. 27.)

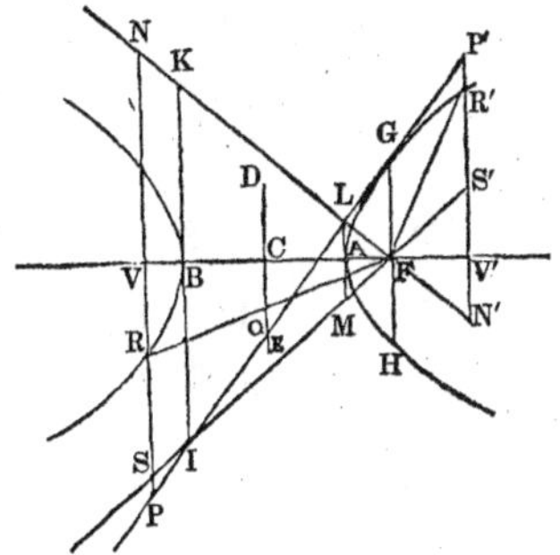

Through the vertices A and B draw, from the focal tangent IT, the lines LA and IB at right angles to AB; through the focus F draw LF and IF, meeting LM and IK in M and K; and produce RV both ways to P and N.

By the principles of construction (15[a]),

IB=BF, and AL=AF.

But[a] IB : BF :: SV : FV, and AL : AF :: VN : FV.

Therefore SV=FV, and VN=FV.

By similar triangles (LGF and LPN) we have

GF : PN :: LG : LP :: [b]AF : AV;

and by similar triangles (GIF and PIS) we have

GF : PS :: GI : PI :: [b]FB : VB.

---

[a] Leg. 4. 18. Euc. 6. 4. [b] Leg. 4. 15, Cor. 2. Euc. 6. 2.

Multiplying the proportions together,

$$GF^2 : PN.PS :: AF.FB : AV.VB.$$

But $PS = PV - SV =$ (by 2) $FR - SV = FR - FV$;

and $PN = PV + VN =$ (by 2) $FR + VN = FR + FV$.

Therefore $PN.PS = \overline{FR+FV}.\overline{FR-FV} =$ [a]$FR^2 - FV^2 =$ [b]RV^2.

Substituting this value of PN.PS, we have

$$GF^2 : RV^2 :: AF.FB : AV.VB.$$

By using the accented letters, P′, R′, S′, V′, N′, the foregoing demonstration will apply to the other branch of the hyperbola.

(80) *Cor.* 1. Hence, if two ordinates are equally distant from the centre, they are equal to one another.

That is, if $CF = CV$, then $GF = RV$.

(81) *Cor.* 2. Hence, the two portions of the curve lying on the opposite sides of the transverse axis AB, or the conjugate axis DE, are similar; and if placed upon one another, would coincide in every part. For if at any point they should not coincide, the ordinates at that point would be unequal, which by Cor. 1 is impossible.

(82) *Cor.* 3. Hence, there is another point situated, in respect to the curve, precisely like the focus F, and may, therefore, be called another focus. Thus, if $CV = CF$, the point V is the other focus.

(Fig. 28.)

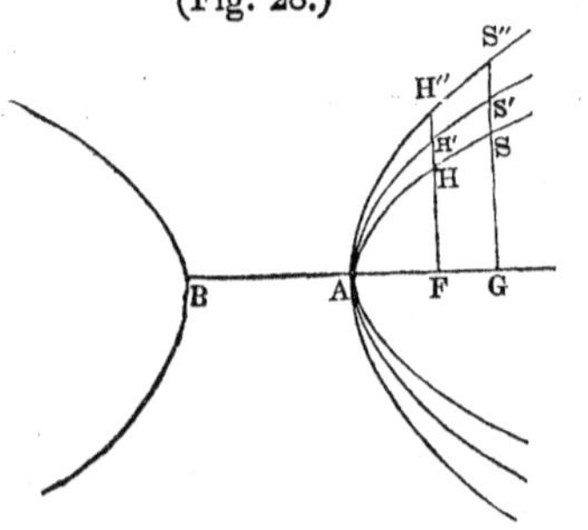

(83) *Cor.* 4. If different hyperbolas have the same transverse axis, the corresponding ordinates are proportional to each other. That is,

$$FH : FH' :: GS : GS'.$$

---

[a] Leg. 4. 10. Euc. 2. 5, Cor. [b] Leg. 4. 11. Euc. 1. 47.

For (79)	$FH^2 : GS^2 :: AF.FB : AG.GB.$
Also,	$FH^2 : GS'^2 :: AF.FB : AG.GB.$
Therefore	$FH^2 : FH'^2 :: GS^2 : GS'^2.$
And[a]	$FH : FH' :: GS : GS'.$

(84) *Cor.* 5. It follows from the last of the foregoing propositions,[b] that $HH' : SS' :: FH : GS$.

(85) *Cor.* 6. Since OC is midway between AL and BI, lying on opposite sides of AB, it equals half their difference,

But $BI - AL = BF - AF = AB$.

Therefore OC is equal to AC, the semi-transverse axis.

(86) Prop. II. Theorem.

The square of an ordinate to the transverse axis, is to the rectangle of the corresponding abscissas, as the square of the conjugate axis is to the square of the transverse axis.

That is, $RV^2 : AV.VB :: DE^2 : AB^2$.

(Fig. 29.)

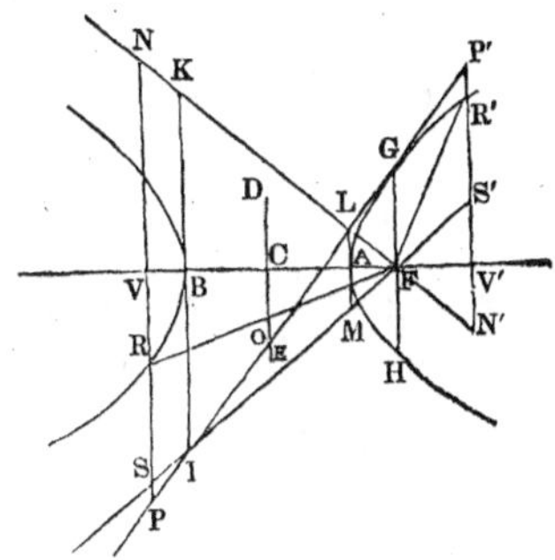

For by similar triangles (IGF and ILM) we have

$$GF : LM = 2AF :: IG : IL :: FB : AB.$$

And by similar triangles (LGF and LIK) we have

$$GF : IK = 2FB :: LG : IL :: AF : AB.$$

---

[a] Leg. 2. 12, Cor. [b] Leg. 2. 6. Euc. 5, D. and 16.

Multiplying the proportions together,

$$GF^2 : 4AF.FB = (7)\ DE^2 :: AF.FB : AB^2.$$

But (79) $GF^2 : RV^2 :: AF.FB : AV.VB.$

Therefore,[a] $RV^2 : AV.VB :: DE^2 : AB^2.$

(87) *Cor.* 1. The line GH is the parameter of the transverse axis.

For, as above, $GF^2 : DE^2 :: AF.FB = \frac{1}{4}DE^2 : AB^2.$

Therefore, extracting roots,[b]

$$AB : \tfrac{1}{2}DE :: DE : GF = \tfrac{1}{2}GH.$$

Or, doubling the second and fourth terms,

$$AB : DE :: DE : GH.$$

(88) *Cor.* 2. If a circle be described on the transverse axis of an hyperbola, an ordinate to the hyperbola is to a tangent to the circle drawn from the foot of the ordinate, as the conjugate axis is to the transverse.

(Fig. 30.)

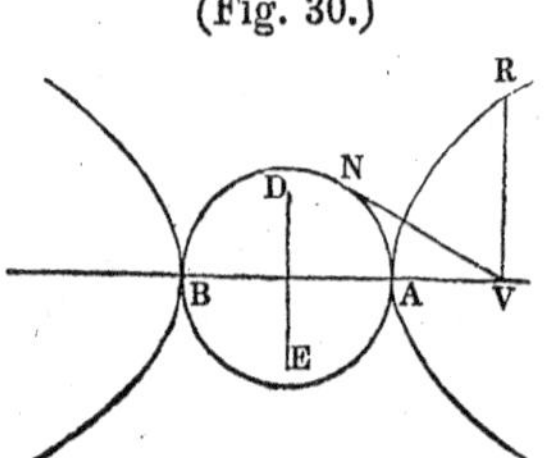

For (86)

$$RV^2 : AV.VB = {}^{c}VN^2 :: DE^2 : AB^2.$$

Therefore, ${}^{b}RV : VN :: DE : AB.$

(89) *Cor.* 3. If the conjugate axis of an hyperbola is equal to the transverse, the hyperbola is said to be *equilateral,* and the square of the ordinate becomes equal to the rectangle of the corresponding abscissas.

(89*a*) *Cor.* 4. $CF^2 - CA^2 = CD^2.$

For ${}^{d}CF^2 - CA^2 = AF.FB = (7)\ CD^2.$

---

[a] Leg. 2. 4. Euc. 5. 24.
[b] Leg. 2. 12, Cor.
[c] Leg. 4 30
[d] Leg. 4. 10. Euc. 2. 5, Cor.

(90) PROP. III. THEOREM.

The difference of two lines drawn from the foci of an hyperbola to any point in the curve, is equal to the transverse axis.

That is, VM−FM=AB.

(Fig. 31.)

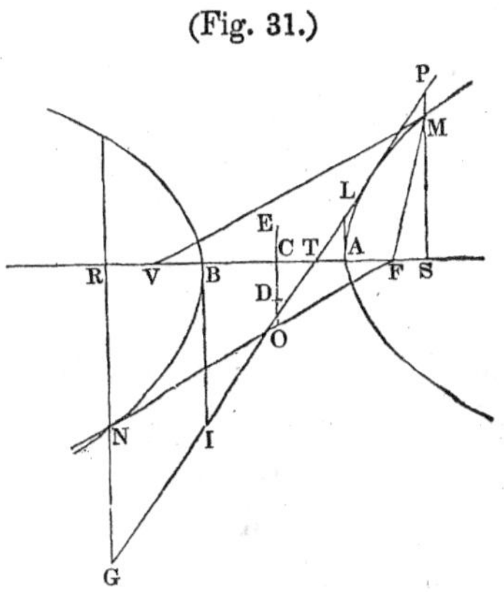

Make the arc BN equal to AM, join FN, draw the ordinates NR and MS, the conjugate axis DE, and the focal tangent GTP, and produce NR, MS, and CD to G, P, and O.

Then, since CO is midway between GR and PS, lying on opposite sides of AB, it is equal to half their difference.

Therefore GR−PS=2OC=(85) AB.
But (2) GR−PS=FN−FM=(81) VM−FM
Therefore VM−FM=AB.

(91) *Scholium.* The property proved in this proposition furnishes the definition of the hyperbola in many treatises. It also affords a ready method of describing the curve mechanically. Take a thread and a ruler, such that the excess of the length of the ruler over that of the thread shall be equal to the transverse axis, and the sum of their lengths greater than the distance between the foci. Fasten one end of each together, and the other ends one to each focus. Place a pencil against the thread, and press it against the ruler so as to keep it constantly stretched, while the ruler is turned around the focus to which it is attached as a centre. The point of the pencil will describe one branch of the hyperbola. For in every position of the pencil, the difference of the distances to the foci will be equal to the difference between the length of the ruler and that of the string.

(92) Prop. IV. Theorem.

Two lines drawn from the foci to any point in the curve, make equal angles with a tangent to the curve at that point.

That is, FM and VM make equal angles with the tangent TU, or FMT=VMT.

If not, let them make equal angles with RS, so that FMR =VMR.

(Fig. 32.)

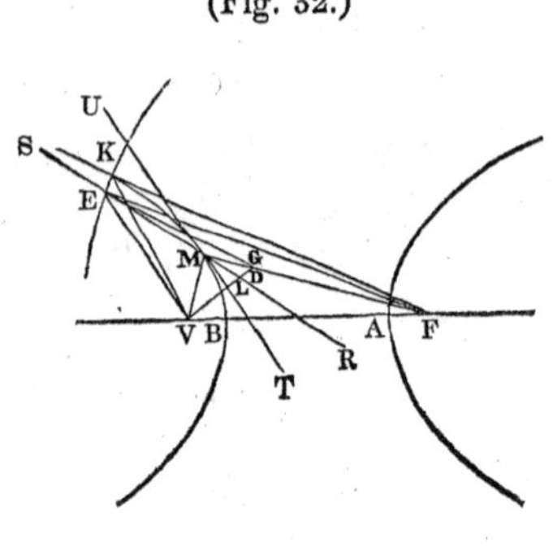

Since there cannot be two different tangents to a curve at the same point, RS must cut it, and fall within, as at some point E. With the centre F and radius FE describe the arc EK, cutting the curve in K. Cut off MG=MV, and join GV, GE, EF, EV, KF, and KV.

The angle RMF=(by supposition) VMR. Then, in the triangles GMD and VMD, GM=MV, and MD is common, and the angle GMD=VMD; therefore GD=VD, and the angle GDM=VDM. Hence, in the triangles EGD and EVD, the sides GD and ED= VD and ED, and the angle GDE=VDE; therefore EG=EV. But EV is less than KV, because the angle EFV is less than KFV, while the sides EF and FV are equal to KF and FV.[a] Therefore EG is less than KV, and consequently EF—EG is greater than KF—KV. Now (90) KF—KV=FM—MV=FM—MG=FG. Hence EF—EG is greater than FG. Or, adding EG to both, EF is greater than FG+EG. That is, one side of a triangle is greater than the sum of the other two sides, which is impossible.[b] In the same manner it may be shown, that no other line but TU makes equal angles with FM and MV, and consequently TU does.

---

[a] Leg. 1. 9. Euc. 1. 24. [b] Leg. 1. 7. Euc. 1. 20.

(93) *Cor* 1. Hence, to draw a tangent to the curve at any point M, join MF and MV, and bisect the included angle VMF.

(94) *Cor.* 2. GV is perpendicular to TU, and is bisected at the point L.

(95) *Cor.* 3. FG=AB, since each is equal to FM−MV.

(96) Prop. V. Theorem.

If a line be drawn from either focus perpendicular to a tangent to the curve at any point, the distance of its intersection from the centre is equal to the semi-transverse axis.

That is, CL=AC.

(Fig. 33.)

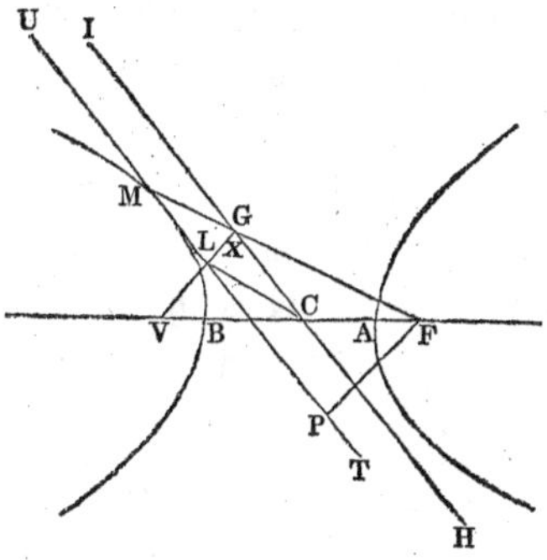

Since FV is bisected in C, and (by the preceding proposition) GV in L, CL is parallel[a] to FG, and the triangles LCV and GFV are similar. Hence

CV : FV :: CL : GF.

But $CV=\frac{1}{2}FV$,

Therefore $CL=\frac{1}{2}GF=(95)\ \frac{1}{2}AB=AC$.

(97) *Cor.* 1. Hence, a circle described on the transverse axis with the centre C, will pass through the intersections L and P; and conversely, if from any point in the circumference of such a circle, two lines be drawn at right angles to one another; and if one of them pass through one of the foci, the other will be a tangent to the curve.

---

[a] Leg. 4. 16. Euc. 6. 2.

(98) *Cor.* 2. Hence, a diameter HI, parallel to TU, would cut off a part MX of the line MF, equal to the semi-transverse axis.

For[a] MX=CL=AC.

(99) Prop. VI. Theorem.

The rectangle of the two perpendiculars drawn from the foci to the tangent to the curve at any point, is equal to the square of the semi-conjugate axis.

That is, $VL.FP=CD^2$.

(Fig. 34.)

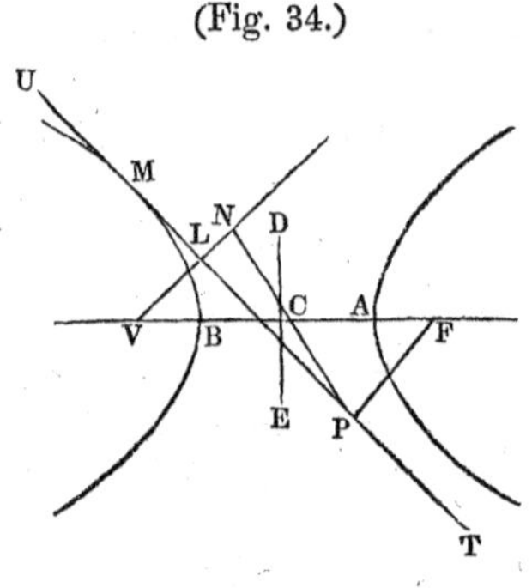

Join PC, and produce PC and LV till they meet in N.

Then will the triangles PFC and NVC be similar and equal, since the angle PCF=NCV, and FPC= the alternate angle VNC, and the side FC=VC. Therefore CN=CP, and (97) the point N is in the circumference of a circle described on AB as a diameter. Consequently,[b] NV.VL=AV.VB. Or, since NV=PF, $PF.VL=AV.VB=(7)\ CD^2$.

(100) Prop. VII. Theorem.

If at any point in the curve a tangent and ordinate be drawn, meeting either axis produced, half that axis is the mean proportional between the distances of the two intersections from the centre.

That is, CS : CA :: CA : CT.

Or, CG : CD :: CD : CH.

---

[a] Leg. 1. 28. Euc. 1. 34. [b] Leg. 4. 29, Cor. Euc. 3. 36.

(Fig. 35.)

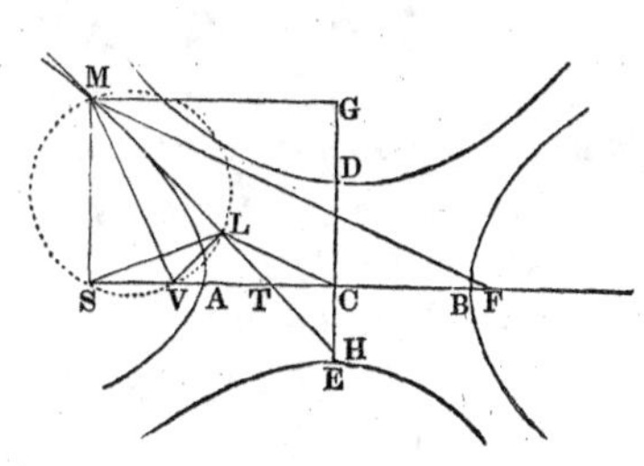

Connect M with the foci F and V, draw VL perpendicular to the tangent, and join CL and SL.

Since MSV and MLV are both right angles, each is an angle in a semicircle,[a] and consequently, a circle described on MV as a diameter, would pass through L and S. Then must the angles VML and VSL be equal, being in the same segment,[b] or measured by the same arc.

But (92) VML = FML = CLT, since CL is parallel to FM. Therefore the angle VSL or CSL=CLT. Hence, the triangles LCT and SCL are similar, for the angle CSL=CLT, and the angle at C is common.

Therefore CS : CL :: CL : CT.

But (96) CL=CA.

Therefore CS : CA :: CA : CT,

which proves the proposition in respect to the transverse axis.

Again, since when three numbers are in continued proportion, the first is to the third as the square of either antecedent is to the square of its consequent, we have from the last proportion

$CS^2 : CA^2 :: CS=GM : CT ::$ (sim. tri.) $GH : CH$.

Hence, by division,[c] $CS^2-CA^2 : CA^2 :: CG : CH$.

But (86) $AS.SB=$[d]$CS^2-CA^2 : CA^2 :: MS^2=CG^2 : CD^2$.

Therefore, by equality of ratios,

$CG : CH :: CG^2 : CD^2$;

which gives, $CG.CH=CD^2$.

Or, $CG : CD :: CD : CH$.

---

[a] Leg. 3. 18, Cor. 2. Euc. 3. 31.
[b] Leg. 3. 18. Euc. 3. 21.
[c] Leg. 2. 6. Euc. 5. 17.
[d] Leg. 4. 10. Euc. 2. 5, Cor.

(101) Prop. VIII. Theorem.

If different hyperbolas have the same transverse axis, the corresponding sub-tangents are equal to one another.

(Fig. 36.)

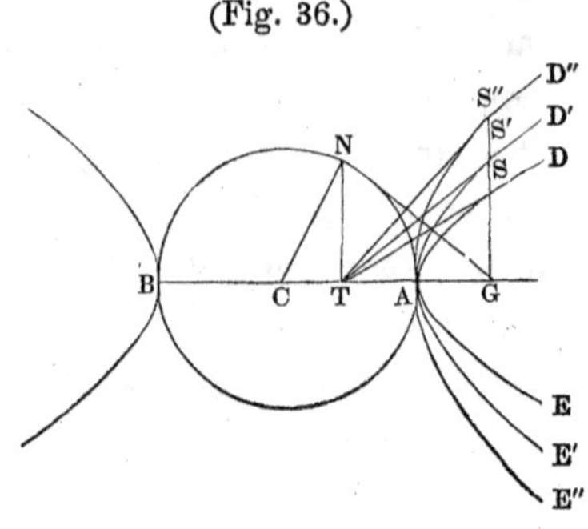

Let EAD, E′AD′, &c. be any number of hyperbolas, each having AB for its transverse axis, SG produced any ordinate to each, and ST, S′T, &c. tangents at the points where the ordinate meets the curves.

Then for each of them we have (100)

$$CG : CA :: CA : CT,$$

and since the first three terms are the same for all, the fourth must be likewise, which, subtracted from CG, gives the sub-tangent.

(102) *Cor.* 1. Hence, we may draw a tangent at a given point in the curve, without knowing the foci.

Let S be the given point. On AB describe a circle; draw the ordinate SG and the tangent to the circle GN, let fall the perpendicular NT, and join TS.

For in the right-angled triangle CNG, we have

$$CG : CN :: CN : CT.$$

But CN=CA; therefore CG : CA :: CA : CT, which is the same proportion as the one above, showing that the ordinate at N and the tangent at S meet the transverse axis in the same point.

(103) *Cor.* 2. CG.GT=AG.GB, each being equal[a] to NG^2.

[a] Leg. 4. 23 and 30. Euc. 6. 8, Cor. and 3. 36.

(104) PROP. IX. THEOREM.

The difference of the squares of two ordinates drawn to either axis from the extremities of any two conjugate diameters, is equal to the square of half the other axis.

That is, $FU^2-GP^2=AC^2$, and $PS^2-UR^2=CE^2$.

(Fig. 37.)

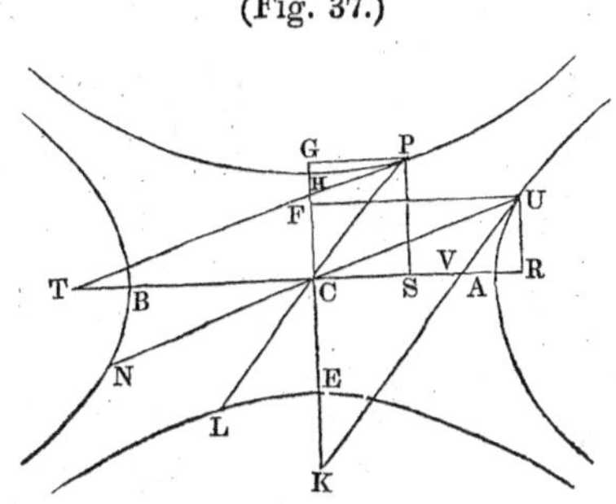

Draw the tangents PT and UK, meeting the transverse axis in T and V, and the conjugate axis in H and K. Then CS.CT= CR.CV, each being equal (100) to AC^2 or BC^2.

Therefore[a] CS : CR :: CV : CT.

But, since (9) UV is parallel to PC, and UC to PT, the triangles PTC and UCV are similar; as also PCS and UVR.

Hence VR : CS :: UV : PC :: CV : CT.

Then, by equality of ratios, we have

CS : CR :: VR : CS,

And, consequently, $CR.VR=CS^2$.

But (103) $CR.VR=AR.BR=$[b]CR^2-AC^2.

Therefore $CR^2-AC^2=CS^2$; or, $CR^2-CS^2=AC^2$.

That is, $FU^2-GP^2=AC^2$.

---

[a] Leg. 2. 2. Euc. 6. 16. [b] Leg. 4. 10. Euc. 2. 5, Cor.

By comparing the similar triangles UKC and PCH, and also UCF and PHG, it may be proved in the same manner that $PS^2-UR^2=CE^2$.

(105) Prop. X. Theorem.

The difference of the squares of any pair of conjugate diameters of an hyperbola is equal to the difference of the squares of the two axes.

That is, $UN^2-PL^2=AB^2-DE^2$.

(Fig. 37*a*.)

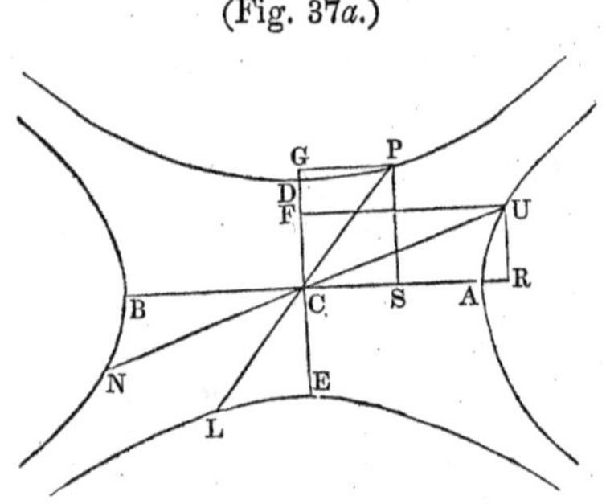

For[a] $UC^2-PC^2=CR^2-CS^2+UR^2-PS^2=CR^2-CS^2-(PS^2-UR^2)$.

But (104) $CR^2-CS^2=AC^2$, and $PS^2-UR^2=CE^2$.

Therefore $UC^2-PC^2=AC^2-CE^2$.

Or, $UN^2-PL^2=AB^2-DE^2$.

---

[a] Leg. 4. 11. Euc. 1. 47.

(106) Prop. XI. Theorem.

Any parallelogram inscribed between conjugate hyperbolas, having its sides parallel to two conjugate diameters, is equal to the rectangle of the two axes.

That is, GHIK=AB.DE.

(Fig. 38.)

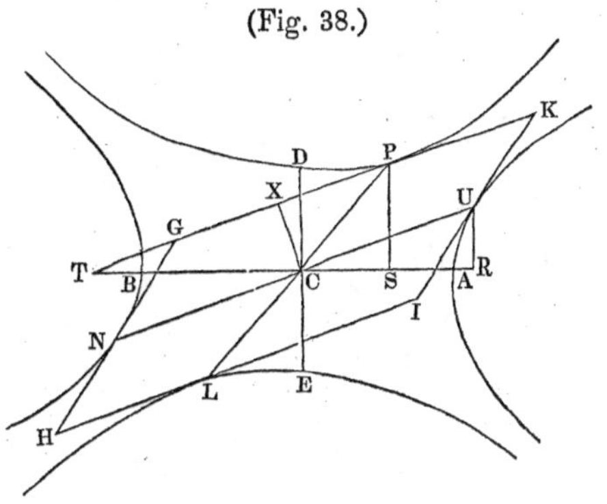

Draw CX at right angles to TK; then will the triangles TCX and CUR be similar.

Now $DC^2 : AC^2 :: UR^2 : AR.RB =$ (as shown in 104) CS^2.

On extracting roots, $DC : AC :: UR : CS$.

Or, alternately, $CS : AC :: UR : DC$.

But (100) $CS : AC :: AC : CT$.

Therefore $UR : DC :: AC : CT$,

or $UR.CT = AC.DC = \frac{1}{4}AB.DE$.

Also (sim. tri.) $UR : CU :: CX : CT$,

or $UR.CT = CU.CX$[a] $= \frac{1}{4}GHIK$.

Therefore GHIK=AB.DE.

(107) *Cor.* $AC : CX :: CU : DC$, or $CU.CX = AC.DC$.

---

[a] Leg. 4. 5.

(108) Prop. XII. Theorem.

The rectangle of two lines, drawn from the foci to any point in the curve, is equal to the square of half the diameter conjugate to that which passes through the point.

That is, $FM \times VM = CH^2$.

(Fig. 39.)

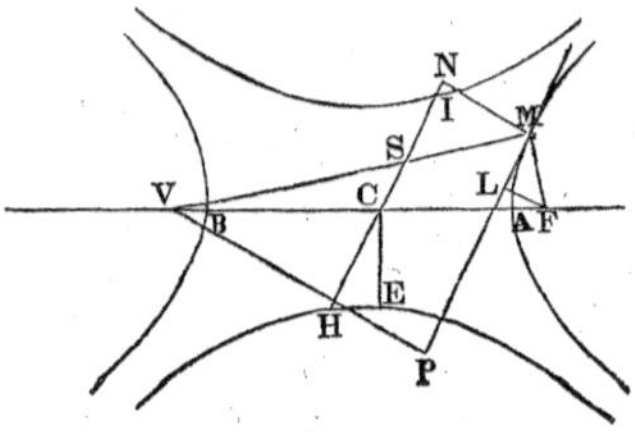

Draw the tangent MLP, and the perpendiculars to it FL, MN, and VP. Then the triangles PMV, SMN, and FML will be similar (92).

Therefore

$MS : MN :: FM : FL$.

And also,

$MS : MN :: VM : VP$.

Multiplying the proportions together,

$$MS^2 : MN^2 :: FM \times VM : FL \times VP.$$

But (98) $MS^2 = AC^2$, and (99) $FL \times VP = CE^2$.

Therefore $AC^2 : MN^2 :: FM \times VM : CE^2$.

Now (107) $MN.CH = AC.CE$.

Therefore $AC : MN :: CH : CE$.

Or, squaring, $AC^2 : MN^2 :: CH^2 : CE^2$.

Hence, by equality of ratios, $CH^2 : CE^2 :: FM \times VM : CE^2$.

And $FM \times VM = CH^2$.

(109) Prop. XIII. Theorem.

If at one of the vertices of an hyperbola a tangent be drawn meeting any diameter, and also from the same point an ordinate to that diameter produced, the semi-diameter is the mean proportional between the distances of the two intersections from the centre.

That is, CE : CI :: CI : CP.

(Fig. 40.)

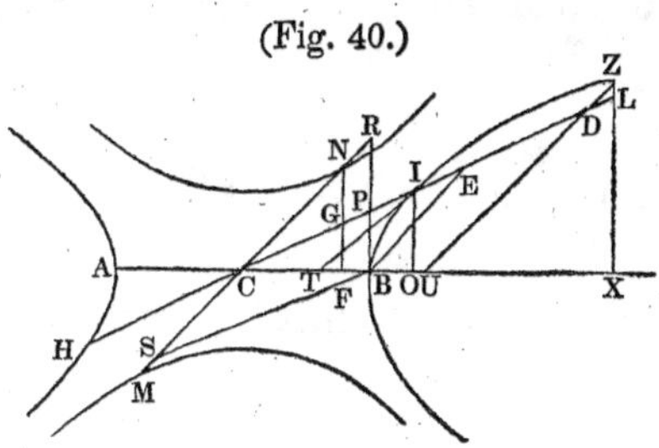

Draw IT and IO tangent and ordinate to the curve at I.

Then (sim. tri.) CE : CI :: CB : CT :: (100) CO : CB :: (sim. tri.) CI : CP.

(110) *Cor.* 1. Hence $CI^2 : CP^2$ :: CE : CP. Also, if the ordinate BS be drawn parallel to HI, we have $CN^2 : CR^2$ :: CS : CR.

For (sim. tri.) $CN^2 : CR^2$ :: CF^2=(as shown in 103) CO.OT : CB^2=(100) CO.CT :: OT : CT :: (111) EP : CP :: (sim. tri.) BE =CS : CR.

(111) *Cor.* 2. The lines CE and CO are similarly divided, the former in I and P, and the latter in B and T; and, consequently, lines joining EO, BI, and PT, would be parallel.

(112) *Cor.* 3. The triangle CIT=CBP, and CEB=COI; for the angle at C is common, and the sides about it reciprocally proportional.[a]

(113) *Cor.* 4. The area of triangle CNG=CBP or CIT.

For[b] CNG : CRP :: $CN^2 : CR^2$ :: (Cor. 1) CS=BE : CR :: (sim. tri.) BP : PR :: [c]CBP : CRP.

Hence, since CNG and CBP have the same ratio to CRP, they are equal to one another.

---

[a] Leg. 4. 24, Cor. Euc. 6. 15. [b] Leg. 4. 25. Euc. 6. 19.
[c] Leg. 4. 6, Cor. Euc. 6. 1.

(114) *Cor.* 5. IOBP=IOT.

For (112) CIT=CBP, and taking each from CIO, the remainders must be equal.

(115) *Cor.* 6. The triangle UZX=PBXL, (Z being any point in the curve.)

For CB : BP :: CX : XL :: [a]CB+CX=AX : BP+XL.

Also, CB : BP :: CO : OI :: CB+CO=AO : BP+OI.

Therefore AX : AO :: BP+XL : BP+OI.

Or,[b] AX.XB : AO.OB :: $\overline{\text{BP+XL}}$.XB : $\overline{\text{BP+OI}}$.OB.

But

$\overline{\text{BP+XL}}$.XB=[c]2PLXB, and $\overline{\text{BP+OI}}$.OB=2PIOB=(114) 2IOT.

Therefore

PLXB : IOT :: AX.XB : AO.OB :: (79) ZX^2 : OI^2 :: [d]UZX : IOT.

Hence, PLXB=UZX, since both have the same ratio to IOT.

(116) *Cor.* 7. DZL=DUC−ICT.

For (112) ICT=CBP=CLX−PLXB=(115) CLX−UZX= DUC−DZL, since the part LDUX is common to both CLX and UZX.

Hence, DZL=DUC−ICT.

(117) Prop. XIV. Theorem.

The square of any diameter is to the square of its conjugate, as the rectangle of its abscissas is to the square of the corresponding ordinate.

That is, HI^2 : MN^2 :: HD.DI : DZ^2.

---

[a] Leg. 2. 10. Euc. 5. 12.
[b] Leg. 2. 8. Euc. 5. 15 and 16.
[c] Leg. 4. 7.
[d] Leg. 4. 25. Euc. 6. 19.

(Fig. 41.)

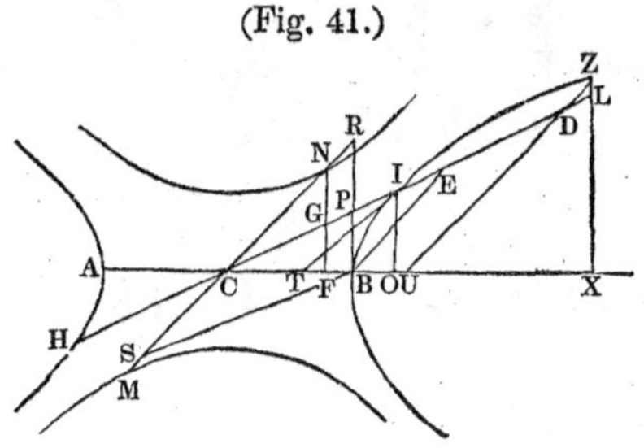

For[a] $CN^2 : DZ^2 :: CNG = (113)\ ICT : DZL = (116)\ DUC - ICT$.

But[a] $ICT : DUC :: CI^2 : CD^2$.

Hence,[b] $ICT : DUC - ICT :: CI^2 : CD^2 - CI^2 =$ [c]$HD.DI$.

Therefore, by equality of ratios,

$$CN^2 : DZ^2 :: CI^2 : HD.DI.$$

Or, alternately, $CI^2 : CN^2 :: HD.DI : DZ^2$.

Or,[d] $HI^2 : MN^2 :: HD.DI : DZ^2$.

(118) *Cor.* Hence, all chords parallel to any diameter are bisected by its conjugate; and, conversely, a line bisecting two or more parallel chords is a diameter.

(119) PROP. XV. PROBLEM.

To draw a tangent to an hyperbola from a given point without the curve.

Let AMBM′ be the given hyperbola, AB the transverse axis, F one of the foci, and T the given point.

Join TF, and upon it and AB as diameters, describe the circles TPF and APB, cutting each other in P and P′. The lines TPM and P′TM′ drawn through the points of intersection, will be tangents to the hyperbola.

---

[a] Leg. 4. 25. Euc. 6. 19. [b] Leg. 2. 6. Euc. 5, D.
[c] Leg. 4. 10. Euc. 2. 5, Cor. [d] Leg. 2. 7. Euc. 5. 15.

Join FP. The angle FPT is[a] a right angle, and FP perpendicular to TM. Now, since from the point P, in the circumference of the circle described on the transverse axis, there are drawn two lines, PF and PM, at right angles to one another, and one of them (PF) passes through the focus, the other must (97) be tangent to the hyperbola.

(Fig. 42.)

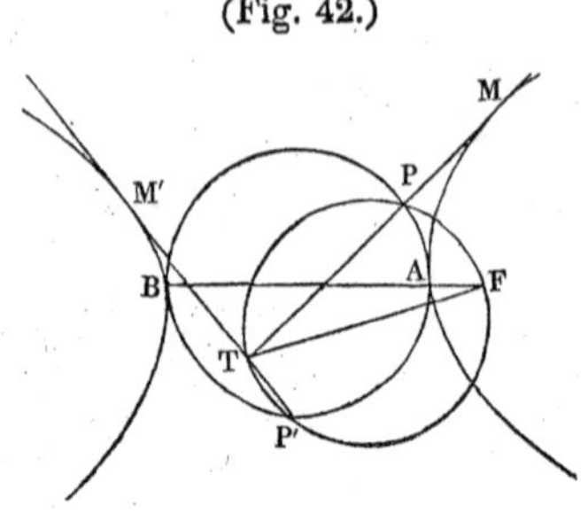

(120) Prop. XVI. Problem.

To find the centre, axes, and foci of a given hyperbola.

(Fig. 43.)

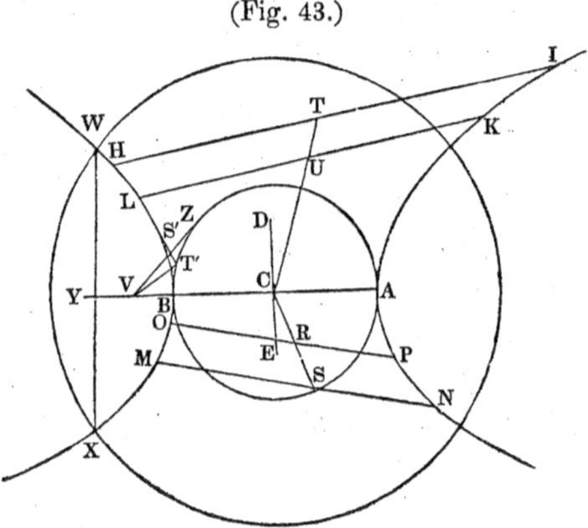

Let AKBL be the given hyperbola.

Draw any two pairs of parallel chords HI and LK, MN and OP; bisect them in T, U, R, and S; join TU and RS, and produce the lines till they meet in C. Both these lines being (118) diameters, the point C must be the centre.

From the centre C, and with any convenient radius, describe a

[a] Leg. 3. 18, Cor. 2. Euc. 3. 31.

circle cutting the hyperbola in any points W and X. Join WX, and at right angles to it draw, through the point C, the line ABY. Since this line bisects [a]WX at right angles, it divides the curve into two similar parts, and therefore (81) AB is the transverse axis.

To find the foci, draw a tangent at any point S′ in the hyperbola, and from the point T′, where it intersects a circle described on the transverse axis, draw T′V perpendicular to it. The point V is (97) one of the foci. From V draw VZ tangent to the circle AT′B, and from C draw CD and CE at right angles to AB, making each equal to VZ. The tangent VZ is [b]the mean proportional between AV and VB, therefore (7) DE is the conjugate axis.

(120*a*) PROP. XVII. THEOREM.

An hyperbola may be formed by the mutual intersection of a cone and a plane.

Let CDLE and CM′U′N′ represent the two nappes of a cone, and CDE and CM′N′ two triangular sections formed by a plane perpendicular to the base, and passing through the vertex of the cone. Let the hyperbola in Prop. I., with no change of letters, be placed one branch upon each nappe, in the manner of a collar, with its plane perpendicular to that of the triangular sections, and the vertices A and B in contact with the surface of the cone somewhere on the lines CM′ and CD. Further, let the base of the cone be so broad that if AB were placed perpendicular to it, the hyperbolas would fall *within* the cone.

(Fig. 44.)

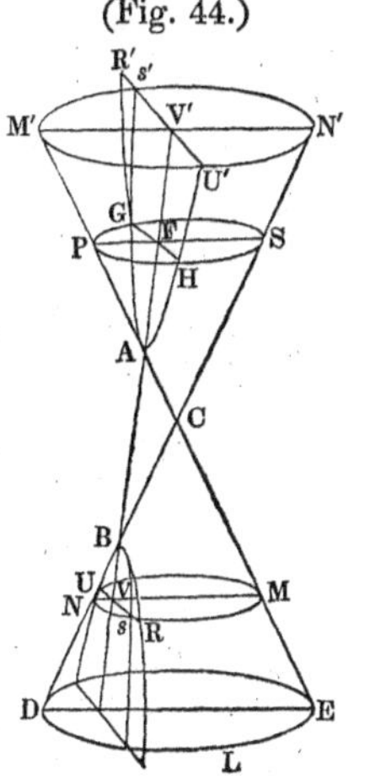

Now, suppose the point A to slide up or down on the line CM′, and B on CD,

[a] Leg. 3. 6. Euc. 3. 3. [b] Leg. 4. 30. Euc. 3. 50.

till the point G shall lie in the surface of the cone; a condition which we shall see to be possible, if we consider that when A or B coincide with C, the hyperbolas must fall wholly *without* the cone. We assert that any other point, R or R′, in the hyperbola will also lie in the surface of the cone.

If not, it must lie either within or without the cone. Let it be supposed to lie without, and that the ordinate RV cuts the surface of the cone at *s*. Through G and *s* let the circular sections PGSH and M*s*NU be made to pass, parallel to the base, and cutting the triangular sections in PS and MN. The lines GF and RV being perpendicular to the plane PSCDE, must also be perpendicular to PS and MN.

By sim. tri. AFP and AVM, PF : MV :: AF : AV.

And by sim. tri. BFS and BVN, FS : VN :: FB : VB.

Multiplying the proportions together,

PF.FS : MV.VN :: AF.FB : AV.VB.

But[a] PF.FS $=$ GF^2, and MV.VN $=$ *s*V^2.

Therefore GF^2 : *s*V^2 :: AF.FB : AV.VB.

But (79) GF^2 : RV^2 :: AF.FB : AV.VB.

Therefore *s*V^2 $=$ RV^2, and *s*V $=$ RV, which is impossible.

In the same manner it may be shown, that the point R cannot lie within the cone, and, consequently, it lies in the surface. And since R is any point in the hyperbola, the whole curve must lie in the surface of the cone.

By using the accented letters M′, R′, *s*′, V′, and N′, the foregoing demonstration will apply to the other branch of the hyperbola.

---

[a] Leg. 4. 23, Cor. Euc. 6. 13.

CHAPTER V.

OF THE CURVATURE OF THE CONIC SECTIONS.

(121) Before entering upon the subject discussed in this chapter, it is necessary to acquaint the student with the doctrine of *ultimate* or *limiting ratios;* a method of investigation much used for determining the ratio between quantities that are not commensurable with each other. For example, if we wish to compare the area of a square with that of its inscribed circle, we may first compare it with the area of a regular polygon inscribed in the circle; and since the greater the number of sides of this polygon the nearer will its area approach to equality with that of the circle, it is assumed that by increasing them indefinitely the two areas will ultimately become equal, or that each will bear the same ratio to the area of the square.[a] This operation involves the following principle, the truth of which will be assumed in the discussions of this chapter, viz.:

The ratio between two quantities is not appreciably[b] *affected by adding to, or subtracting from either, an indefinitely small part of itself.*

If a grain of sand were annihilated, it would hardly affect the ratio which the weight of the whole earth bears to that of the moon, or any other body; but even this would be far greater than in the cases in which we employ limiting ratios.

---

[a] Leg. 5. 8. Euc. Sup. 1. 4.

[b] Strictly it is not affected at all; for in the limit these indefinitely small quantities are really nothing. We employ the language for the sake of convenience.

(122) *Def.* The *curvature* of a curve is its deviation from a tangent line, measured by the subtense of an indefinitely small arc. Thus the curvatures of the two curves AG and AE at the point A, are to each other as the subtenses BD and BC of the indefinitely small arcs AD and AC.

(Fig. 45.)

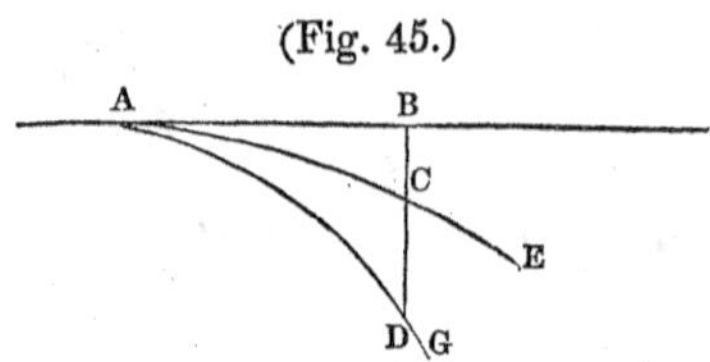

(123) *Def.* If in any curve three points be taken at equal distances, but indefinitely near each other, the circle which passes through them [a]is called an *osculating circle*, and through the indefinitely small arcs lying between those points, the two curves may be considered to coincide, that is, to touch one another. And further, since curves may be regarded as polygons of an indefinite number of sides, the parts of the curves lying between contiguous points thus taken may be considered straight lines.

(124) *Def.* The radius of the osculating circle is called the *radius of curvature* of the curve at the point of contact, and its diameter the *diameter of curvature.* Also any chord that passes through the point of contact is called a *chord of curvature.*

(125) Prop. I. Theorem.

The radius of curvature at the vertex of a conic section, is equal half the principal parameter.

Let GAH be the conic section, AB its transverse axis, OMN the osculating circle at the vertex A, and AM an indefinitely small arc common to both curves.

Put the parameter equal to P, and draw the ordinate MS.

---

[a] Leg. 3. 7. Euc. 4. 5.

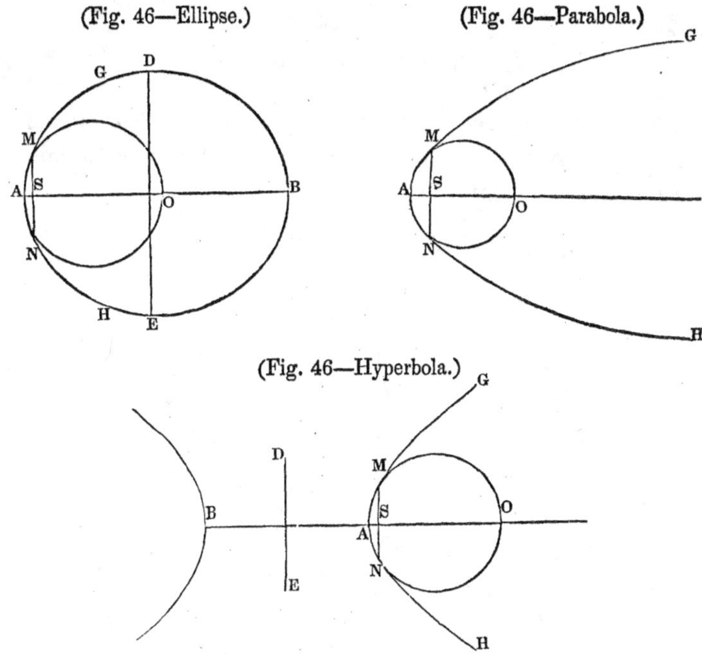

(Fig. 46—Ellipse.) (Fig. 46—Parabola.)

(Fig. 46—Hyperbola.)

In the ellipse and hyperbola (12),

	AB : DE :: DE : P, the parameter.
Hence (38*b*),	$AB^2 : DE^2$ :: AB : P.
But (24 and 86),	$AB^2 : DE^2$ :: AS.SB : SM^2=[a]AS.SO.
Therefore,	AB : P :: AS.SB : AS.SO :: SB : SO.
Or, alternately,	AB : SB :: P : SO.

Now, the nearer the point M is to A, the nearer do the lines SB and SO approach to equality with AB and AO, and in the limit at A the ratio between them becomes that of equality (121). The last proportion will then read AB : AB :: P : AO. Consequently, P=AO, or $\frac{1}{2}$P=$\frac{1}{2}$AO, which proves the truth of our proposition in regard to the ellipse and hyperbola.

Again, in the parabola, we have (60) AS.P=SM^2=AS.SO.

Therefore P=SO=(in the limit at A) AO, or $\frac{1}{2}$P=$\frac{1}{2}$AO.

---

[a] Leg. 4. 23, Cor. Euc. 6. 13.

(126) Prop. II. Theorem.

In the ellipse and hyperbola the chord of curvature that passes through the centre is equal to the parameter of the diameter that passes through the point of contact.

That is, ML : HI :: HI : MO.

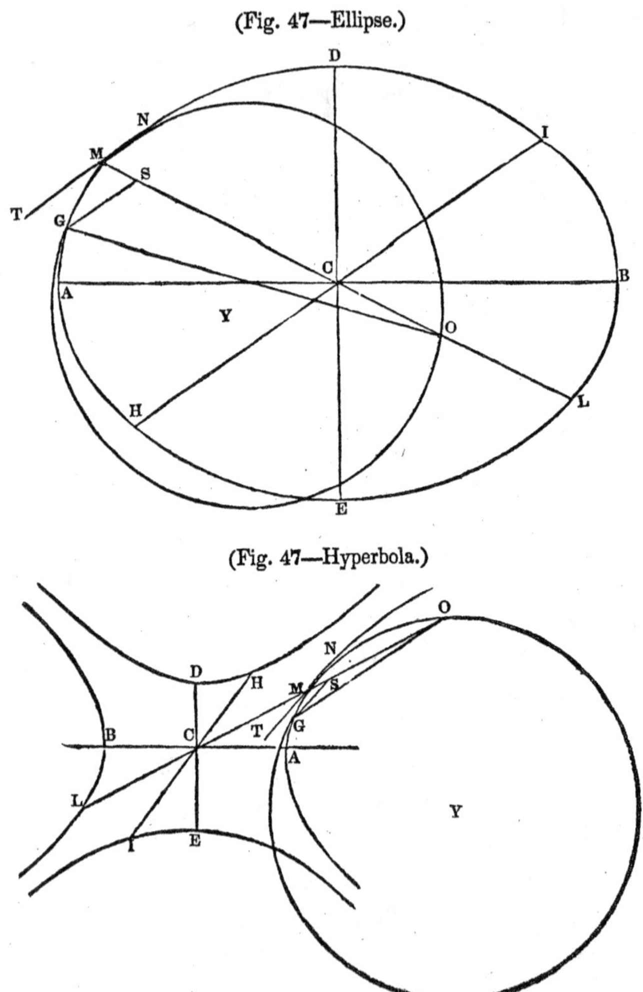

(Fig. 47—Ellipse.)

(Fig. 47—Hyperbola.)

Let N, M, and G be the three points through which the osculating circle OMG is drawn (123). Join MC and produce it to O; join OG, produce NM to T, and from G draw GS parallel to MT.

The triangles MGS and MGO are similar, for the angle at M is common, and SGM=GMT=[a]MOG. Therefore

MO : MG :: MG : MS.

And, since by the definition of an osculating circle (123), MG is but an indefinitely small part of MO, MS must be but an indefinitely small part of MG, and much more then must MS be but an indefinitely small part of MO. Consequently (121) the ratio of MO to SO, and also that of ML to SL, is to be regarded as the ratio of equality. Moreover, since the arc MG is indefinitely small, the chords MO and OG are to be regarded as equally distant from Y the centre of the circle, and therefore [b]equal to one another. But MO : OG :: MG : SG; therefore the ratio of MG to SG, or of MG^2 to SG^2, is that of equality.

By (54) and (117), ML^2 : HI^2 :: MS.SL : SG^2; and, dividing the first and third terms by the equals ML and SL, and multiplying them by MO, we have ML.MO : HI^2 :: MS.MO=(by the first of our proportions above) MG^2 : SG^2. But the ratio of MG^2 to SG^2 is that of equality, therefore ML.MO=HI^2; or,

ML : HI :: HI : MO.

(127) *Cor.* ML.MO=HI^2, or MC.MO=$2CH^2$.

(128) Prop. III. Theorem.

In the parabola the chord of curvature that passes through the focus is equal to the parameter of the diameter that passes through the point of contact.

---

[a] Leg. 3. 21. Euc. 3. 32. [b] Leg. 3. 8. Euc. 3. 14.

That is, MS : SG :: SG : MR.

(Fig. 48.)

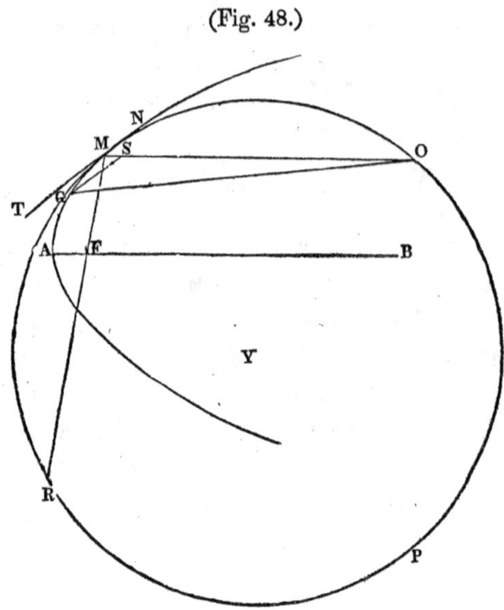

Let N, M, and G be the three points through which the osculating circle OMG is drawn (123). Draw MO parallel to the transverse axis AB, join OG, produce NM to T, and draw GS parallel to MT.

It may be proved, as in (126), that MS : MG :: MG : MO, and also that in the limit MG=SG; therefore MS : SG :: SG : MO. But since (61) MO and MR make equal angles with the tangent TN, they cut off equal arcs of the circle,[a] and are therefore themselves equal.

Therefore MS : SG :: SG : MR.

(129) *Cor.* MR=4FM. For (76*a*) MS : SG :: SG : 4FM.

---

[a] Leg. 3. 21. Euc. 3. 32 or 34.

(130) Prop. IV. Theorem.

In the ellipse and hyperbola the chord of curvature that passes through the focus is a third proportional to the transverse axis, and a diameter conjugate to that which passes through the point of contact.

That is, AB : HI :: HI : MR, drawn through the focus F

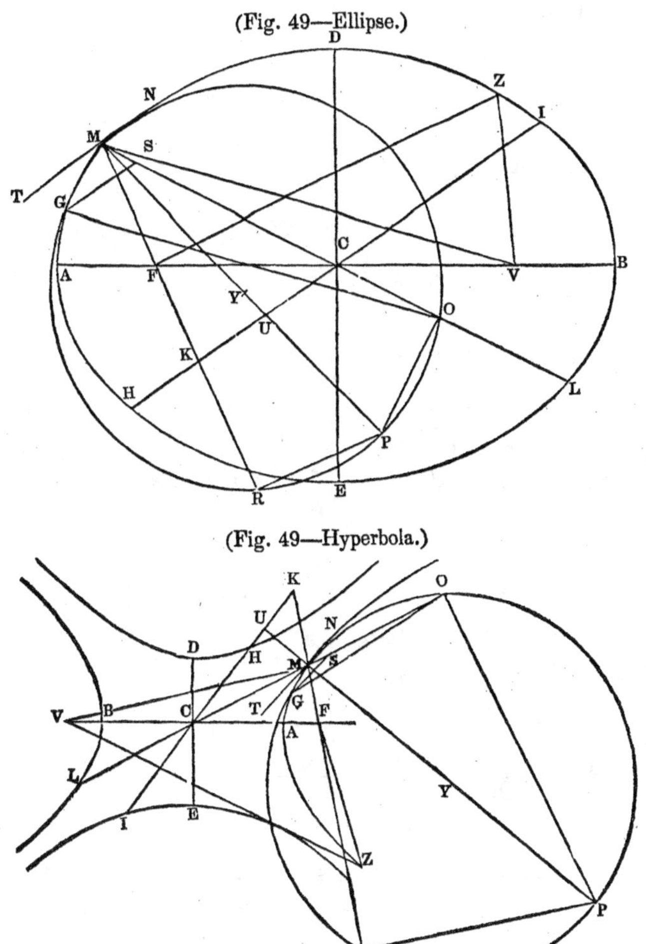

(Fig. 49—Ellipse.)

(Fig. 49—Hyperbola.)

Draw MP through the centre Y of the osculating circle, and consequently at right angles to TM and HI; and join OP and RP. The angles MRP and MOP are [a]right angles, and consequently the triangles MCU and MOP are similar, as also MKU and MRP.

Hence, OM : MP :: MU : MC, and OM.MC=MP.MU.

Also, MR : MP :: MU : MK=(36 and 98) AC,
and MR.AC=MP.MU.

Consequently, MR.AC=OM.MC,
and OM : MR :: AC : MC :: AB : ML.
But (126) OM : HI :: HI : ML.
Therefore AB : HI :: HI : MR.

(131) *Cor.* 1. $MR=\frac{HI^2}{AB}$.

(131*a*) *Cor.* 2. We have above $MP.MU=OM.MC=(127)\ 2CH^2$.
Hence, $MY.MU=CH^2$.

(132) Prop. V. Theorem.

In the ellipse and hyperbola the squares of the radii of curvature at different points of the curve, are to each other as the cubes of the rectangles of the distances of each from the two foci.

That is (Fig. 49), putting R and *r* for the radii of curvature at M and Z

$$R^2 : r^2 :: \overline{FM.MV}^3 : \overline{FZ.ZV}^3.$$

For (44 and 107), CH.MU=AC.DC.

And (131*a*), $MY.MU=CH^2$.

Therefore $AC.DC : CH^2 :: CH : MY=R$.

Hence $R=\frac{CH^3}{AC.DC}$, and $R^2=\frac{CH^6}{(AC.CD)^2}$.

But (45 and 108), $FM.MV=CH^2$.

And, consequently, $\overline{FM.MV}^3=CH^6$.

---

[a] Leg. 3. 18, Cor. 2. Euc. 3. 31.

Therefore $R^2=\frac{\overline{FM.MV}^3}{(AC.DC)^2}$.

In the same manner it may be shown that $r^2=\frac{\overline{FZ.ZV}^3}{(AC.DC)^2}$.

Therefore $R^2 : r^2 :: \overline{FM.MV}^3 : \overline{FZ.ZV}^3$.

(132*a*) *Cor.* The radius of curvature varies as the cube of the diameter conjugate to that which passes through the point of contact. For, in the equation $R=\frac{CH^3}{AC.DC}$, the denominator of the fraction is constant; and therefore R varies as CH^3.

(133) PROP. VI. THEOREM.

In the parabola the squares of the radii of curvature at different points of the curve, are to each other as the cubes of the distances from the focus.

That is, putting R and r for the radii of curvature at M and Z,

$$R^2 : r^2 :: FM^3 : FZ^3.$$

(Fig. 51.)

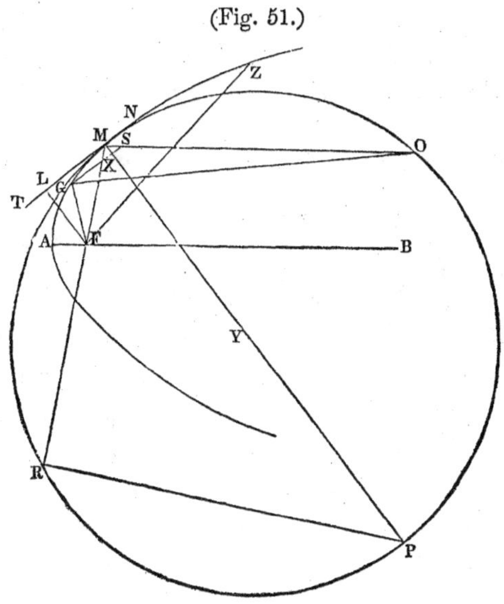

Draw MP the diameter of the osculating circle, join RP, and draw FL perpendicular to MT. Then will the triangles MRP and FML be similar.[a]

Hence $\quad MP : MR = (129)\ 4FM :: FM : FL.$

And squaring, $\quad MP^2 : \overline{4FM}^2 :: FM^2 : FL^2 = (68c)\ AF.FM.$

Therefore $\quad MP^2 : \overline{4FM}^2 :: FM : AF,$ and $MP^2 = \frac{16\overline{FM}^3}{AF}.$

In the same manner it may be shown that the square of the diameter of curvature at $Z = \frac{16\overline{FZ}^3}{AF}$. Therefore the square of the diameter of curvature at M : the square of the diameter of curvature at Z :: $FM^3 : FZ^3$, and the radii will have the same ratio;

That is, $\quad R^2 : r^2 :: FM^3 : FZ^3.$

(134) Prop. VII. Theorem.

If straight lines be drawn from one of the foci of a conic section to the curve, so as to cut off indefinitely small but equal sectors, the curvatures of the included arcs towards that focus are to each other inversely as the square of their distances from it.

That is, $Mx : Aw :: AF^2 : FM^2$, provided the areas FAO and FMG are indefinitely small but equal.

First, in the case of the ellipse and hyperbola.[b] Let AB and DE be the axes, F and V the foci, M any point in the curve, and KAO and MGRP osculating circles at the points A and M.

It may be shown, in the same manner as in the first proportion in (126), that $Mx : MG :: MG : MR$, and consequently that $Mx.MR = MG^2$; and also that $Aw.AK = AO^2$.

[a] Leg. 3. 21 and 18, Cor. 2. Euc. 3. 31 and 32.

[b] It is not thought necessary to add the figure for the hyperbola, as it is perfectly analogous to that for the ellipse.

(Fig. 52—Ellipse.)

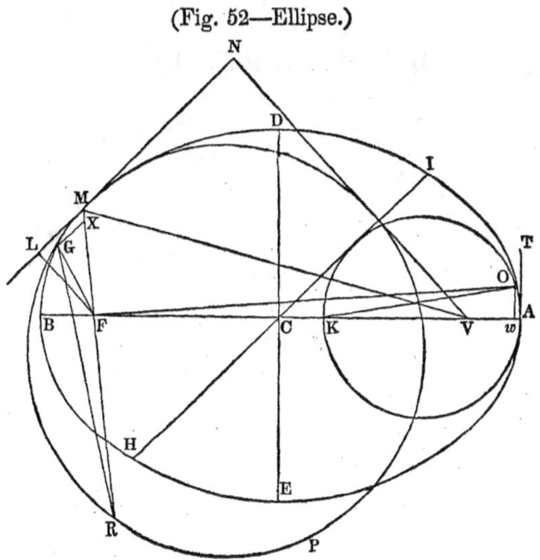

Therefore $Mx.MR : Aw.AK :: MG^2 : AO^2$.

But (131), $MR=\frac{HI^2}{AB}=\frac{4CH^2}{AB}=$(45 and 108) $\frac{4FM.MV}{AB}$

And (125), $AK=\frac{DE^2}{AB}=$(38 and 99) $\frac{4FL.VN}{AB}$.

By substituting these values into the proportion above, we readily obtain

$$Mx.FM.MV : Aw.FL.VN :: MG^2 : AO^2.$$

But again, since in the limit LM coincides with MG, and AT with AO, FL becomes the altitude of the triangle GFM, and AF of FAO, and the areas of these triangles being equal, we have[a]

$$AF.AO=FL.MG.$$

Or, $AF : FL :: MG : AO$.

And squaring, $AF^2 : FL^2 :: MG^2 : AO^2$.

Hence, by equality of ratios and dividing by FL,

$$Mx.FM.MV : Aw.VN :: AF^2 : FL.$$

[a] Leg. 4. 6.

By the similar triangles, MNV and MLF (30 and 92), we have the proportion,

$$\text{MV} : \text{VN} :: \text{FM} : \text{FL}.$$

By which divide the preceding one, and we get

$$\text{M}x.\text{FM} : \text{A}w :: \frac{\text{AF}^2}{\text{FM}} : 1.$$

Dividing the first and third by FM, and multiplying the third and fourth by FM^2,

$$\text{M}x : \text{A}w :: \text{AF}^2 : \text{FM}^2.$$

(Fig. 52—Parabola.)

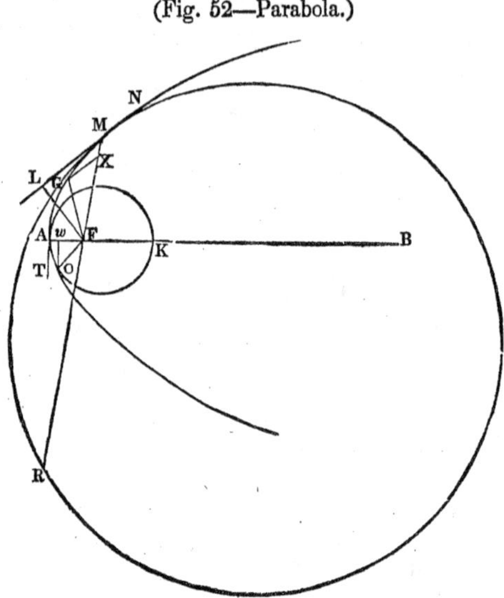

(134*a*) Secondly, in the parabola it may be shown, in the same manner as in the ellipse and hyperbola, that

$$\text{M}x.\text{MR} : \text{A}w.\text{AK} :: \text{MG}^2 : \text{AO}^2.$$

And also that $\text{AF}^2 : \text{FL}^2 :: \text{MG}^2 : \text{AO}^2.$

Therefore, by equality of ratios,

$$\text{M}x.\text{MR} : \text{A}w.\text{AK} :: \text{AF}^2 : \text{FL}^2.$$

But (129) MR=4FM, and (125 and 59) AK=4AF, and (68*c*) FL^2=AF.FM. Hence, by substitution,

$$Mx.FM : Aw.AF :: AF^2 : AF.FM.$$

Dividing the second and fourth by AF, and transferring the factor FM from the first to the fourth,

$$Mx : Aw :: AF^2 : FM^2.$$

(135) *Schol.* On the property proved in this proposition depends the important law that the paths of the heavenly bodies, and of all others under the influence of gravitation, are necessarily conic sections.

CHAPTER VI.

OF SOME PROPERTIES PECULIAR TO THE DIFFERENT CONIC SECTIONS.

---

(136) Prop. I. Theorem.

The area of a parabola is equal to two-thirds of the circumscribing parallelogram.

That is, $MAP=\frac{2}{3}MBLP$.

(Fig. 53.)

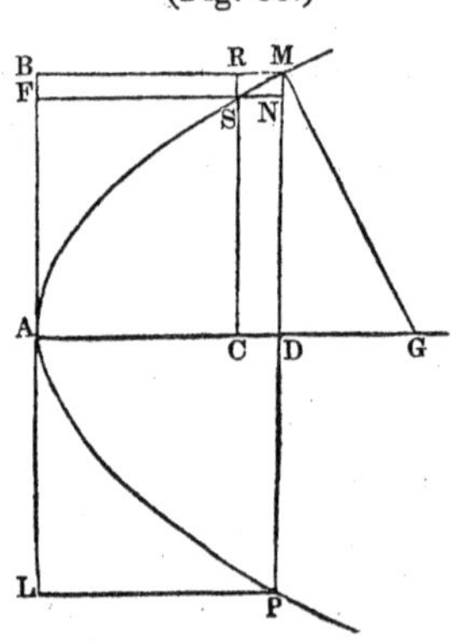

Through the point S indefinitely near to M draw NF parallel to the abscissa AD, and RC parallel to the ordinate MD; also draw the normal MG. The triangles MNS and MDG are similar,[a] since the sides of the one are respectively perpendicular to those of the other. Therefore

$$MN : SN :: DG : MD.$$

Or, multiplying the first and third terms by $AD=MB$, and the second and fourth by MD,

$$MN.MB : SN.MD :: AD.DG : MD^2 = (60 \text{ and } 68b)\, 2AD.DG :: 1 : 2.$$

That is, the interior rectangle MC is double the external one MF, and the same would be true of any other rectangles similarly drawn. Consequently the whole space AMD is double of ABM, and hence equal to $\frac{2}{3}ABMD$; or $MAP=\frac{2}{3}MBLP$.

---

[a] Leg. 4. 21.

(137) Prop. II. Theorem.

The area of an ellipse is to that of a circle described on its transverse axis, as the conjugate axis is to the transverse.

That is, ADBE : AD‴BE‴ :: DE : AB.

(Fig. 54.)

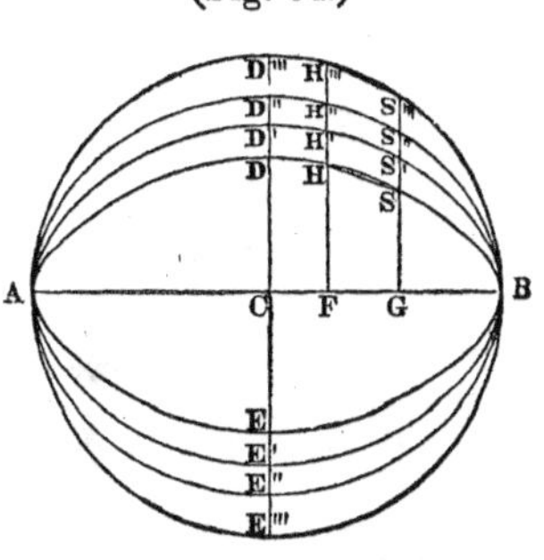

Draw any two ordinates H‴F and S‴G, indefinitely near each other.

It follows readily from (21)[a], that the area of HFGS : the area of H‴FGS‴ :: CD : CD‴ :: DE : AB, and the same would be true of any other trapezoids similarly drawn. We may therefore suppose them indefinitely increased so as to occupy the entire area of the ellipse and circle, and shall then have the area of ADBE : the area of AD‴BE‴ :: DE : AB.

(138) Prop. III. Theorem.

The sum of the first, second, or third powers of four lines drawn from one of the foci of an ellipse to the extremities of any pair of conjugate diameters is the same, whatever may be the position of those diameters.

That is, $VI + VD + VH + VP =$ a constant quantity.

Or, $VI^2+VD^2+VH^2+VP^2=$ a constant quantity.

Or, $VI^3+VD^3+VH^3+VP^3=$ a constant quantity.

Join FP and FI.

Put the semi-transverse axis$=a$.

" the semi-conjugate axis$=b$.

" the eccentricity (14)$=c$.

" $VI=x$, VH or $FI=y$, VD or $FP=x'$, and $VP=y'$.

[a] Leg. 4. 7.

(Fig. 55.)

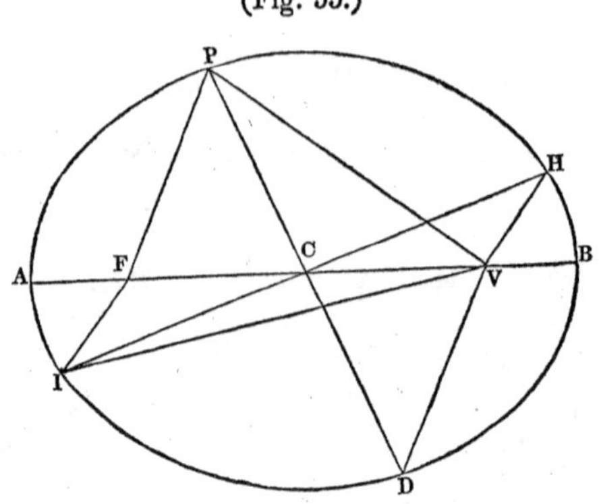

CASE 1.

By (28) $x+y=2a$, and also $x'+y'=2a$.

Therefore $x+y+x'+y'=4a$, a constant quantity.

(139) *Cor.* 1. Hence the mean value of lines drawn from one of the foci to different points of the curve is equal to the semi-transverse axis.

CASE 2.

By algebra we have $x^2+y^2=(x+y)^2-2xy$.

But (28) $(x+y)^2=4a^2$, and (45) $2xy=2\text{CP}^2$.

Therefore $x^2+y^2=4a^2-2\text{CP}^2$.

In like manner $x'^2+y'^2=4a^2-2\text{CI}^2$.

Therefore $x^2+y^2+x'^2+y'^2=8a^2-2(\text{CP}^2+\text{CI}^2)=$(42) $8a^2-2(a^2+b^2)$ $=6a^2-2b^2=$(27) $4a^2+2c^2$, a constant quantity.

(140) *Cor.* 2. Hence the mean value of the squares of lines drawn from one of the foci of an ellipse to different points of the curve is equal to $a^2+\frac{1}{2}c^2$; that is, to the square of the semi-transverse axis, plus half the square of the eccentricity.

Case 3

By algebra we have

$$x^3+y^3=(x+y)^3-3x^2y-3xy^2=(x+y)^3-3\,(x+y)\,xy.$$

But (28) $x+y=2a$, and hence $(x+y)^3=8a^3$; also (45) $xy=\text{CP}^2$.

Therefore $\quad x^3+y^3=8a^3-6a.\text{CP}^2$.

In like manner $\quad x'^3+y'^3=8a^3-6a.\text{CI}^2$.

Therefore $x^3+y^3+x'^3+y'^3=16a^3-6a\,(\text{CP}^2+\text{CI}^2)=$ (42) $16a^3-6a\,(a^2+b^2)=10a^3-6ab^2=4a^3+6a\,(a^2-b^2)=4a^3+6ac^2$, a constant quantity.

(141) *Cor.* 3. Hence the mean value of the cubes of lines drawn from one of the foci of an ellipse to different points of the curve is equal to $a^3+1\frac{1}{2}ac^2$; that is, to the cube of the semi-transverse axis, plus the square of the eccentricity multiplied by three-fourths of the transverse axis.

(142) Prop. IV. Problem.

To find the mean value of the reciprocals of a given power of lines drawn from one of the foci of an ellipse of small eccentricity to different points of the curve.

Put the semi-transverse axis$=a$.
" the semi-conjugate axis$=b$.
" the eccentricity$=c$.
" VI (Fig. 55)$=a+x$.
" VH or FI$=$(28) $a-x$.
" VP$=a+y$.
" VD or FP$=$(28) $a-y$.
" the index of the given power$=n$.

Then $\dfrac{1}{\text{VI}^n}=\dfrac{1}{(a+x)^n}=$ (by algebra) $\dfrac{1}{a^n}-\dfrac{nx}{1.a^{n+1}}+\dfrac{n\,(n+1)\,x^2}{1.2.a^{n+2}}-\dfrac{n\,(n+1)\,(n+2)\,x^3}{1.2.3.a^{n+3}}$, &c.

And $\frac{1}{VH^n}=\frac{1}{(a-x)^n}=$ (by algebra) $\frac{1}{a^n}+\frac{nx}{1.a^{n+1}}+\frac{n\ (n+1)\ x^2}{1.2.a^{n+2}}+$

$$\frac{n\ (n+1)\ (n+2)\ x^3}{1.2.3.a^{n+3}},\ \&c.$$

By the conditions of the proposition the value of x or y is but a small fraction of a, and hence these series converge so rapidly that it will be sufficiently accurate to employ only the first four terms. By adding the two series together we obtain,

$$\frac{1}{VI^n}+\frac{1}{VH^n}=\frac{2}{a^n}+\frac{n\ (n+1)\ x^2}{a^{n+2}}.$$

In like manner, $\frac{1}{VP^n}+\frac{1}{VD^n}=\frac{2}{a^n}+\frac{n\ (n+1)\ y^2}{a^{n+2}}$.

Hence $\frac{1}{VI^n}+\frac{1}{VH^n}+\frac{1}{VP^n}+\frac{1}{VD^n}=\frac{4}{a^n}+\frac{n\ (n+1)}{a^{n+2}}\ (x^2+y^2)$; and the mean value of the four is $\frac{1}{a^n}+\frac{n\ (n+1)}{4a^{n+2}}\ (x^2+y^2)$.

Now (45) $CP^2=VI.IF=(a+x)\ (a-x)=a^2-x^2$.
And, in like manner, $CH^2=a^2-y^2$.
Therefore $CP^2+CH^2=2a^2-(x^2+y^2)$.
But (42) $CP^2+CH^2=a^2+b^2$, and (27) $b^2=a^2-c^2$.
Therefore $2a^2-(x^2+y^2)=2a^2-c^2$, and $x^2+y^2=c^2$.

By substituting c^2 in the place of (x^2+y^2) in the foregoing expression for the mean value, it becomes $\frac{1}{a^n}+\frac{n\ (n+1)\ c^2}{4a^{n+2}}$, an expression containing only constant quantities; and since HI and DP are any conjugate diameters, it must be true for the entire circumference.

(143) *Schol.* This proposition enables us to find the mean attraction of the sun upon any planet throughout its entire orbit, and would do so equally well if the force of gravity varied inversely as the third, fourth, or any higher power of the distance.[a] By means

[a] That is, if we suppose it possible for the law of gravity to be changed, and the orbit still to retain its elliptical form.

of the principle involved in it Laplace succeeded in discovering the true cause of the *secular acceleration of the moon's mean motion*, a subject which had very much perplexed previous astronomers.

In the case of gravity $n=2$, and the expression for the mean value becomes $\frac{1}{a^2}+\frac{3c^2}{2a^4}$, from which we learn that the attraction is greater than it would be if the planet revolved in a circle at the same mean distance in the ratio $a^2+1\frac{1}{2}c^2 : a^2$.

PART II.

ANALYTICAL GEOMETRY.

CHAPTER I.

OF THE POINT AND STRAIGHT LINE IN A PLANE.

(144) The position of a point in a plane may be determined in either of two ways, viz.: by determining its distances from two given lines in the plane that intersect one another, or by determining its distance and direction from a given point in the plane. Thus, the position of the point P may be determined by knowing its distances PD and PE from the fixed lines AX and AY, in which case it is said to be determined by *rectilineal co-ordinates;* or by knowing the distance AP and the angle PAD, in which case it is said to be determined by *polar co-ordinates.* Hence, in a plane, rectilineal co-ordinates consist of two straight lines, and polar co-ordinates of a line and an angle.

(Fig. 57.)

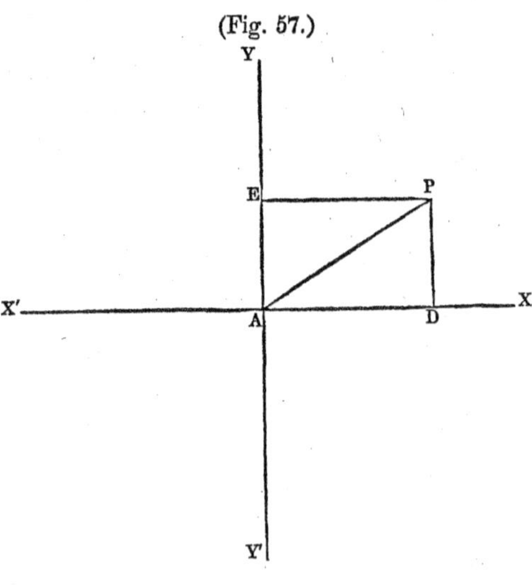

(145) In the former case the fixed lines X′AX and YAY′ are called *co-ordinate axes*, or *axes of reference*, and taken separately the first is called the *axis of abscissas*, and the second the *axis of ordinates*. The point of intersection A is called the *origin*.

(146) The line PE parallel to AX is called the *abscissa* of the point P, and the line PD parallel to AY is called the *ordinate*. Taken together they are called *co-ordinates*, as already remarked. Instead of PE we may employ its equal AD as the abscissa.

(147) If the axes are at right angles to one another the co-ordinates are said to be *rectangular*, but if not they are called *oblique*.

(148) The angle YAX is called the *first* angle, YAX′ the *second*, X′AY′ the *third*, and Y′AX the *fourth*.

(149) All ordinates drawn upwards from X′AX are considered *positive*, and those drawn downwards *negative*; while all abscissas drawn from YAY′ to the *right* are considered positive, and those drawn to the *left* negative. Hence the co-ordinates of a point situated in the first angle are both positive; in the second, the abscissa is negative and the ordinate positive; in the third, they are both negative; and in the fourth, the abscissa is positive, but the ordinate is negative.

(150) It is plain that a single *point* can have but one abscissa, and but one ordinate; but a *line*, since it contains an indefinite number of points, can have an indefinite number of pairs of co-ordinates, varying in their length, and hence spoken of as *variable quantities*. It is customary to denote the abscissa by the letter x, and the ordinate by y.

(151) If polar co-ordinates are employed the point A is called the *pole*, the line AX the *angular axis*, AP the *radius vector*, and PAX the *variable angle*.

In pursuing the subject we shall at first employ rectangular co-ordinates, as these are more simple in their application; but shall show, before closing the chapter, how they may be transformed into oblique or polar ones.

(152) *Def.* *The equation of a line, whether straight or curved, is one that expresses the relation between the co-ordinates of any point in the line.*

For example, if in Fig. 57 PD is two-thirds as long as PE or AD, the same ratio would exist[a] between the co-ordinates of any other point in the line AP; so that wherever they were drawn we should have $y=\frac{2}{3}x$. This is, therefore, the equation of the line AP.

(153) PROP. I. THEOREM.

The general equation of a straight line is

$$y=ax+b;$$

in which a represents the tangent of the angle that the line makes with the axis of abscissas, b the portion of the axis of ordinates intercepted between the line and the origin, and x and y the co-ordinates of any point in the line.

Let AX and AY be the axes, A the origin, PT any straight line, and AD and PD co-ordinates of any point P in the line.

(Fig. 58.)

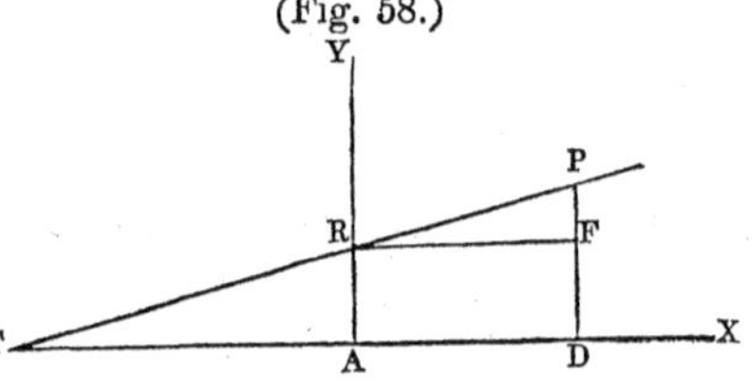

Put AD=x.
" PD=y.
" AR=b.
" tan. PTD=a.

By trigonometry, PF=RF tan. PRF.

But[b] tan. PRF=tan. PTD=a, and RF=AD=x.

Therefore PF=ax.

Also FD=AR=b.

Therefore $ax+b$=PF+FD=PD=y.

Or $y=ax+b$.

---

[a] Leg. 4. 18. Euc. 6. 4. [b] Leg. 1. 20, Cor. 3. Euc. 1. 29.

(154) *Cor.* 1. If the line passes through the origin, $b=o$, and the equation reduces to $y=ax$, a form analogous to that used above as an example, in which $a=\frac{2}{3}$.

(155) *Cor.* 2. If the line is parallel to the axis of abscissas, $a=o$, and the equation reduces to $y=b$. It is evident also, from an inspection of the figure, that in that case every ordinate would be equal to AR.

(156) Prop. II. Theorem.

The equation of a straight line passing through a given point is

$$y'-y=a\ (x'-x);$$

in which x' and y' represent the co-ordinates of the given point, x and y those of any other point in the line, and a the tangent of the angle that the line makes with the axis of abscissas.

Let RT be the straight line, M the given point, and AD and PD the co-ordinates of any other point P in the line.

(Fig. 59.)

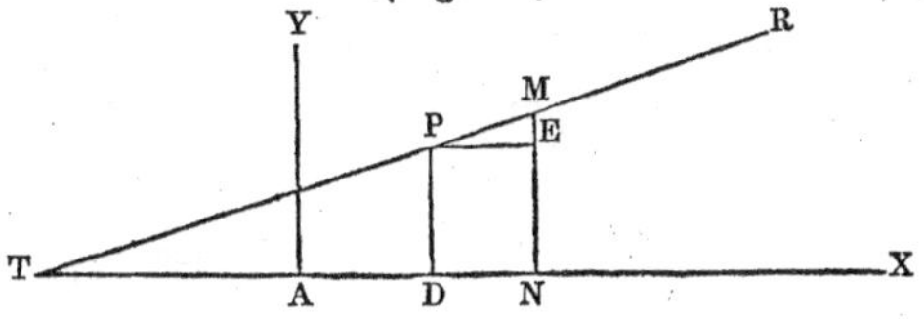

Put AD$=x$.
" PD$=y$.
" AN$=x'$.
" MN$=y'$.
" tan. PTD$=$tan. MPE$=a$.

Then ME$=y'-y$, and PE$=x'-x$.
But ME$=$PE. tan. MPE.
Therefore $y'-y=a\ (x'-x)$.

(156*a*) Otherwise; since Prop. I. is true of every point in TM, we have the two equations,

$$y=ax+b.$$
$$y'=ax'+b.$$

Eliminating b by means of these equations, we have

$$y'-y=a\ (x'-x).$$

(157) PROP. III. THEOREM.

The equation of a straight line passing through two given points is

$$y'-y=\frac{y''-y'}{x''-x'}(x'-x);$$

in which x' and y' represent the co-ordinates of one of the given points, x'' and y'' those of the other, and x and y those of any other point in the line.

Let RT be the straight line, M and R the given points, and AD and PD the co-ordinates of any point P.

Put AD $=x$.
" PD $=y$.
" AN $=x'$.
" MN $=y'$.
" AS $=x''$.
" RS $=y''$.

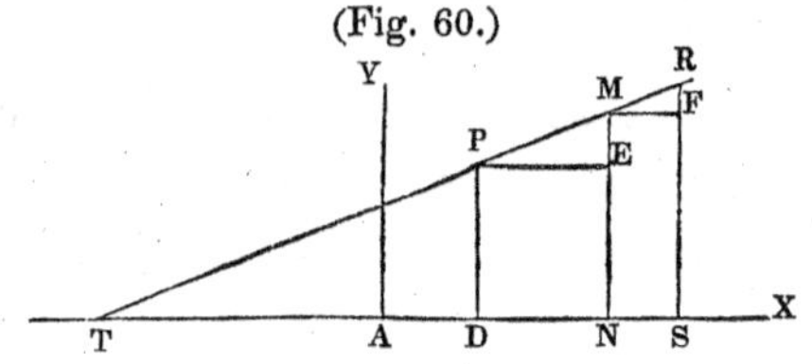

Then ME $=y'-y$, PE $=x'-x$, RF $=y''-y'$, and MF $=x''-x'$.

By sim. tri. MF : RF :: PE : ME.

Hence MF.ME = RF.PE, or, $\text{ME}=\frac{\text{RF}}{\text{MF}}\text{PE}$.

That is, $$y'-y=\frac{y''-y'}{x''-x'}(x'-x).$$

(157*a*) Otherwise; tan. PTS = tan. RMF = (by trig.) $\frac{\text{RF}}{\text{MF}}=\frac{y''-y'}{x''-x'}$, which substitute in the place of a in the last proposition, and we get, as before, $y'-y=\frac{y''-y'}{x''-x'}(x'-x)$.

(157*b*) Otherwise; by Prop. I. we have the three equations,

$$y=ax+b.$$
$$y'=ax'+b.$$
$$y''=ax''+b.$$

Eliminating a and b by means of these three equations, we have

$$y'-y=\frac{y''-y'}{x''-x'}(x'-x).$$

(158) Prop. IV. Theorem.

Every equation of the first degree between two variables is the equation of a straight line.

For, by transposing and uniting terms, every such equation can be reduced to the form $Ay+Bx+C=0$, in which A, B, and C represent any constant quantities, whether positive or negative. But the above equation reduces to $y=-\frac{B}{A}x-\frac{C}{A}$, in which $-\frac{B}{A}$ answers to a in the first proposition, and $-\frac{C}{A}$ to b; that is, it reduces to the form $y=ax+b$.

Example. Prove that the equation $20-x+7y=10x-12$ is the equation of a straight line.

(159) Prop. V. Theorem.

The distance between two points is equal to $\sqrt{(y'-y)^2+(x'-x)^2}$; in which x' and y' are the co-ordinates of one point, and x and y those of the other.

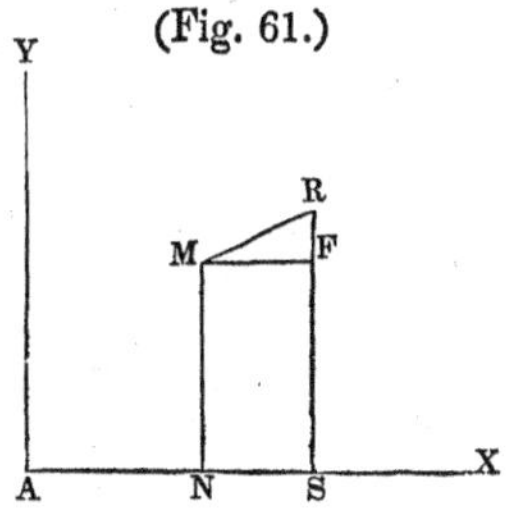

Let M and R be the two points. (Fig. 61.)

Put $AN=x$.

" $MN=y$.

" $AS=x'$.

" $RS=y'$.

Then $RF=y'-y$, and $MF=x'-x$.

But[a] $\overline{MR}^2=\overline{RF}^2+\overline{MF}^2=(y'-y)^2+(x'-x)^2$.

And, extracting the root, $MR=\sqrt{(y'-y)^2+(x'-x)^2}$.

---

[a] Leg. 4. 11. Euc. 1. 47.

(160) Prop. VI. Theorem.

The tangent of the angle included between two straight lines is

$$\frac{a'-a}{1+aa'};$$

in which a represents the tangent of the angle that one of them makes with the axis of abscissas, and a' the other.

Let VR and VS be the two lines intersecting one another in V.

(Fig. 62.)

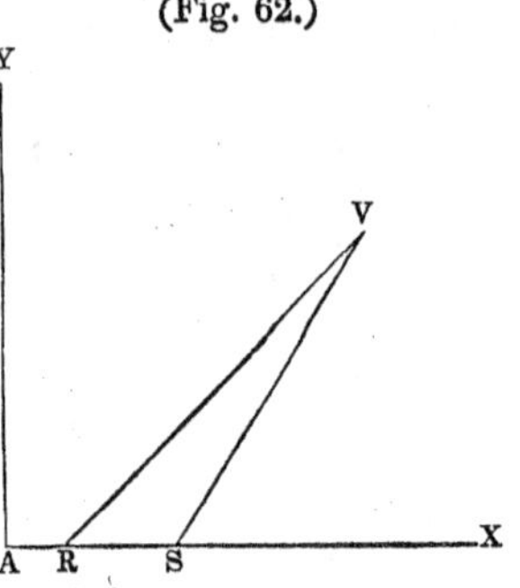

Put tan. VRX$=a$.
" tan. VSX $=a'$.

Because VSX is the exterior angle of the triangle VRS, we have[a] VSX= RVS+VRS, or RVS=VSX−VRX; and consequently, tan. RVS = tan. (VSX − VRX).

But, by trigonometry,[b] the formula for the tangent of the difference of two angles whose tangents are a' and a, is $\frac{a'-a}{1+aa'}$.

Therefore tan. RVS$=\frac{a'-a}{1+aa'}$.

(161) *Cor.* If the lines are parallel $a'-a=o$, and if perpendicular to one another $1+aa'=o$. For when the value of a fraction is nothing its numerator must be nothing, and when the value is infinite the denominator must be nothing.

We have thus far used only rectangular co-ordinates, but it is sometimes more convenient to employ oblique or polar ones. It is, therefore, important to be able to pass from one system to the other; that is, to be able to find the oblique or polar co-ordinates

---

[a] Leg. 1. 25, Cor. 6. Euc. 1. 32. [b] Leg. Trigonometry, Art. XXV.

of a point from the rectangular ones, and the reverse. The process is called

TRANSFORMATION OF CO-ORDINATES.

(162) Prop. VII. Problem.

To pass from a system of rectangular to a system of oblique co ordinates.

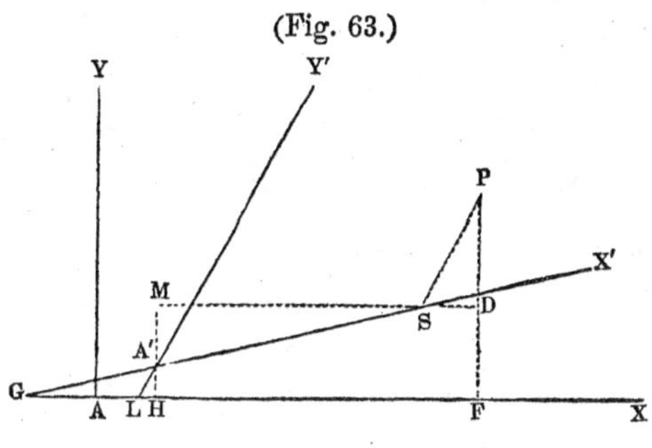

(Fig. 63.)

Let AX and AY be the primitive rectangular axes, and AF and FP the co-ordinates of any point P. Let A′X′ and A′Y′ be the new axes, A′S and PS the new co-ordinates of the point P. Also, let AH and A′H be the co-ordinates of the origin of the new system, and X′GX and Y′LX the angles which the new axes make with the primitive axis of abscissas.

Put AF$=x$, PF$=y$, A′S$=x'$, SP$=y'$, AH$=m$, A′H$=n$, X′GX $=a$, and Y′LX$=a'$.

Then, by trigonometry,

$$\text{MS}=\text{A}'\text{S}.\ \cos.\ \text{A}'\text{SM}=x'\cos.\ a.$$

$$\text{And}\quad \text{SD}=\text{SP}.\ \cos.\ \text{PSD}\ =y'\cos.\ a'.$$

Therefore $x'\cos.\ a+y'\cos.\ a'=\text{MS}+\text{SD}=\text{HF}.$

(162*a*) Or, adding AH, $x'\cos.\ a+y'\cos.\ a'+m=\text{AF}=x.$

Also, by trigonometry, $\text{A}'\text{M}=\text{A}'\text{S}.\ \sin.\ \text{A}'\text{SM}=x'\sin.\ a.$

And, adding A′H, $\text{MH or DF}=x'\sin.\ a+n.$

Also, by trigonometry, $\text{PD}=\text{SP}.\ \sin.\ \text{PSD}=y'\sin.\ a'.$

(162*b*) Hence $x'\sin.\ a+n+y'\sin.\ a'=\text{DF}+\text{PD}=\text{PF}=y.$

Now, if in any equation of a line referred to the primitive axes we substitute for x and y their values just obtained, we shall have the equation for the new axes.

(163) *Cor.* 1. If the origin is the same in both systems, m and n disappear, and the expressions for x and y become

$$x=x' \cos. a+y' \cos. a', \text{ and } y=x' \sin. a+y' \sin. a'.$$

(164) *Cor.* 2. If the new axes are parallel to the primitive ones, $a=0°$, and $a'=90°$, which reduces the expression for x to $x'+m$, and for y to $y'+n$.

(165) *Cor.* 3. If the new axes are rectangular, but not parallel to the primitive ones, $a'=90°+a$, which changes the expression for x to $x' \cos. a-y' \sin. a+m$, and that for y to $x' \sin. a+y' \cos. a+n$.

For then,[a] $\sin. a'=\sin. 90° \cos. a+\cos. 90° \sin. a=\cos. a.$

And $\cos. a'=\cos. 90° \cos. a-\sin. 90° \sin. a=-\sin. a.$

(166) *Schol.* It matters not in which of the four angles formed by the primitive axes the origin of the new ones is placed, provided the proper signs are prefixed to the co-ordinates m and n. And further, since the only effect of these co-ordinates is to add their lengths to the values otherwise obtained, we may, in any case, first find the value of the primitive co-ordinates in terms of the new ones, and then add the co-ordinates of the new origin.

(167) Prop. VIII. Problem.

To pass from a system of oblique to a system of rectangular co-ordinates.

Using the same notation and figure as in the last proposition, and supposing the origin to be the same in both systems, we have (163)

$$x=x' \cos. a+y' \cos. a', \text{ and } y=x' \sin. a+y' \sin. a'.$$

---

[a] Davies' Legendre, Trigonometry, Art. XIX.

Solving these equations for x' and y', and adding (166) the co-ordinates of the new origin, designated by m' and n', we obtain

$$x'=\frac{x \sin. a'-y \cos. a'}{\sin. a' \cos. a-\sin. a \cos. a'}+m'={}^{a}\frac{x \sin. a'-y \cos. a'}{\sin. (a'-a)}+m'.$$

$$y'=\frac{y \cos. a-x \sin. a}{\sin. a' \cos. a-\sin. a \cos. a'}+n'=\frac{y \cos. a-x \sin. a}{\sin. (a'-a)}+n'.$$

Now if, in any equation of a line referred to the primitive oblique axes, we substitute for x' and y' their values just obtained, we shall have the equation for the new rectangular axes.

(168) *Schol.* By the aid of the two preceding propositions we can pass from one system of oblique co-ordinates to another, by first passing from the primitive oblique system to a rectangular one by Prop. VIII., and from that to the new oblique system by Prop. VII.

(169) Prop. IX. Problem.

To pass from a system of rectangular to a system of polar co-ordinates.

(Fig. 65.)

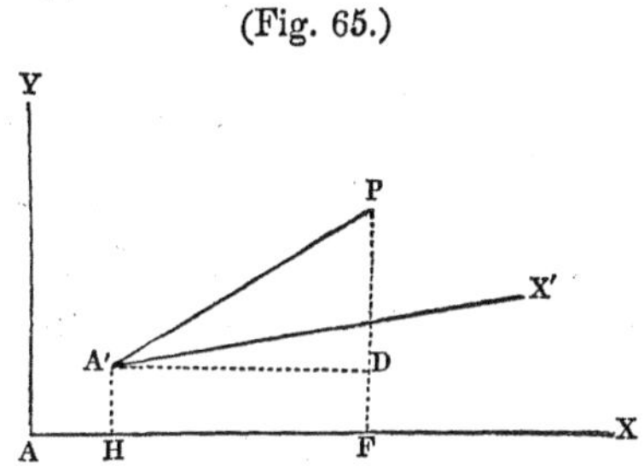

Let AX and AY be the primitive axes, and AF and FP the co-ordinates of the point P.

Let A′X′ be the new angular axis, A′ the pole, AH and A′H its co-ordinates, A′P the radius vector, PA′X′ the variable angle (151), and DA′X′ the angle which the new angular axis makes with the primitive axis of abscissas.

Put $AH=m$, $A'H=n$, $A'P=r$, $DA'X'=a$, $X'A'P=\omega$, $AF=x$ and $PF=y$.

---

[a] Davies' Legendre, Trigonometry, Art. XIX.

Then HF=A D=(by trigonometry) A′P cos. PA′D=r cos. $(a+\omega)$.

(169a) And AF=AH+HF=$m+r$ cos. $(a+\omega)=x$.

Again, by trigonometry, PD=A′P sin. PA′D=r sin. $(a+\omega)$.

(169b) And PF=A′H+PD=$n+r$ sin. $(a+\omega)=y$.

These values of x and y being substituted in any equation in the same manner as in Prop. VII., will give us the polar equation.

CHAPTER II.

OF CURVES.

---

(170) Prop. I. Problem.

To construct a curve from its equation.

Draw the axes X′AX and AY; in X′AX take at pleasure any points, −1, 1, 2, 3, 4, 5, &c., and calculate the corresponding ordinates from the equation of the curve. Lay off these ordinates parallel to AY, and above or below X′AX, according as the value of y is found to be positive or negative, and through their extremities, as a, b, c, d, &c., draw the curve.

(Fig. 66.)

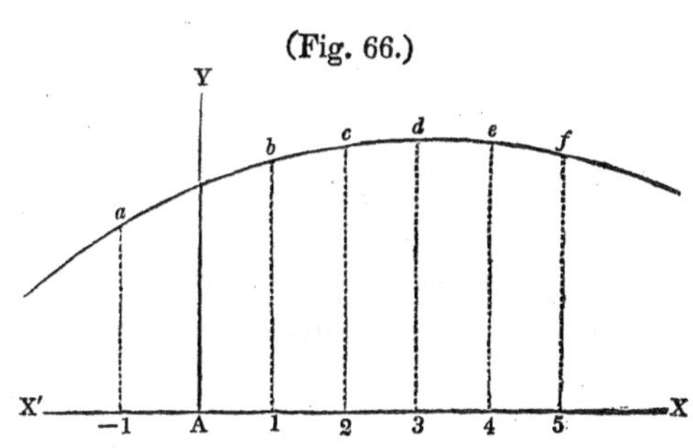

For example, suppose the equation of the curve to be $y=ax^3$, and that $a=5$. Giving arbitrary numerical values to x, (suppose the series of numbers from −4 to +5, as shown below,) and from the equation computing for each value of x the corresponding value of y, we obtain the series set against y.

$$x= \quad -4, \quad -3, \quad -2, \quad -1, \quad 0, \quad 1, \quad 2, \quad 3, \quad 4, \quad 5,$$
$$y=-320, \quad -135, \quad -40, \quad -5, \quad 0, \quad 5, \quad 40, \quad 135, \quad 320, \quad 625.$$

These values, being laid off upon the axes according to their signs (149), will enable us to trace the curve.

(171) *Examples.* Construct the following curves, putting $a=8$, $b=5$, $m=10$, $n=8$, $p=7$, and $r=6$; and taking care in the last six to draw both values of y.

1st. $my=ax^2+b$, giving to x values from -5 to $+5$ inclusive.

2d. $y=x^2+b$, " " " -5 to $+5$ "

3d. $y=bx-a$, " " " -5 to $+5$ "

4th. $(y-n)^2=r^2-(x-m)^2$, giving to x values from $+3$ to $+17$ inclusive.

5th. $y^2=ax^2-x^3+b$, giving to x values from -4 to $+10$ inclusive, and also the value 8,07.

6th. $y^2=r^2-x^2$, giving to x values from -7 to $+7$ inclusive.

7th. $y^2=px$, " " " -2 to $+8$ "

8th. $y^2=\frac{b^2}{a^2}(a^2-x^2)$ " " " -9 to $+9$ "

9th. $y^2=\frac{b^2}{a^2}(x^2-a^2)$ " " " -16 to $+16$ "

(172) Prop. II. Theorem.

The equation of a circle is $(x-m)^2+(y-n)^2-r^2=0$; *in which* m *and* n *denote the co-ordinates of the centre, and* r *the radius.*

(Fig. 67.)

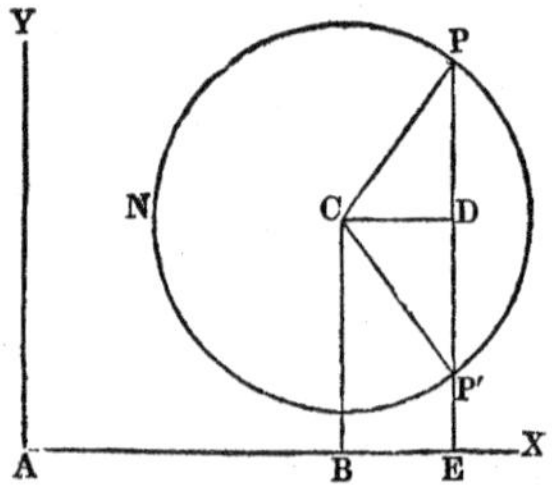

Let PNP′ be a circle, AB and BC the co-ordinates of the centre C, and AE and EP those of any point P in the circumference.

Draw CD parallel to AX, and join CP. Put AB$=m$, BC$=n$, CP$=r$, AE $=x$, and EP$=y$.

Then CD$=$AE$-$AB$=x-m$, and PD$=$EP$-$BC$=n-y$.

But $CD^2+PD^2=CP^2$.

That is, $(x-m)^2+(y-n)^2=r^2$;

or, by transposition, $(x-m)^2+(y-n)^2-r^2=0$.

(173) *Cor.* If the origin is at the centre, m and n disappear, and the equation reduces to $x^2+y^2-r^2=0$, a form often met with.

(174) Prop. III. Problem.

To find the equation of a tangent to a circle, the origin of the co-ordinates being the centre.

(Fig. 68.)

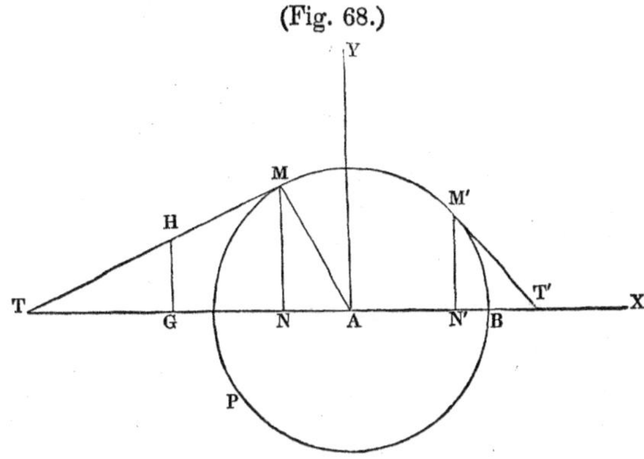

Let MT be a line touching the circle BMP at the point M, it is required to find its equation.

Since it passes through the point M its equation must (156) be of the form $y'-y=a(x'-x)$, in which x' and y' are the co-ordinates of the point M, x and y those of any other point H in the line MT, and a the tabular tangent of the angle MTX, or its equal [a]AMN.

By trig. MN : AN :: R : tan. AMN.

That is, $y' : x' :: 1 : a$.

Hence $a=\frac{x'}{y'}$.

(175) Now of the three values, a, x' and y', that enter into this equation, either all are negative, or one only. Thus, if the point be taken in the first angle, as at M', x' and y' will be positive (149),

[a] Leg. 4. 23. Euc. 6. 8.

but the tangent of M′T′X negative, since the angle is between 90° and 180°.[a] If it be taken in the second angle, as at M, x' will be negative, y' positive, and a positive, since the angle MTX is between 0° and 90°.

In the same manner it may be shown that if the point be taken in the third quadrant, a, x', and y' will all be negative; and in the fourth, a and x' positive, but y' negative. So that the equation will read—

In the 1st quadrant $-a=\frac{x'}{y'}$;

In the 2d " $a=\frac{-x'}{y'}$;

In the 3d " $-a=\frac{-x'}{-y'}$;

In the 4th " $a=\frac{x'}{-y'}$.

All of which can be reduced to the form

(175a) $$a=-\frac{x'}{y'}.$$

Substituting this value of a into the equation $y'-y=a\,(x'-x)$, we obtain for the equation of the tangent MT,

(175b) $$y'-y=-\frac{x'}{y'}\,(x'-x);$$

which may readily be reduced to the form $x'x+y'y-r^2=0$, by clearing of fractions, transposing, and for $x'^2+y'^2$ substituting the equal value r^2. See Appendix B.

[a] It is proved in treatises on trigonometry (see Davies' Legendre, Trigonometry, Art. XII.) that the tangent of an angle is positive when the angle is between 0° and 90°, or between 180° and 270°; but negative when the angle is between 90° and 180°, or between 270° and 360°.

(176) Prop. IV. Problem

To find the polar equation of a circle.

Let P be the pole, and PX′ parallel to AX the angular axis.

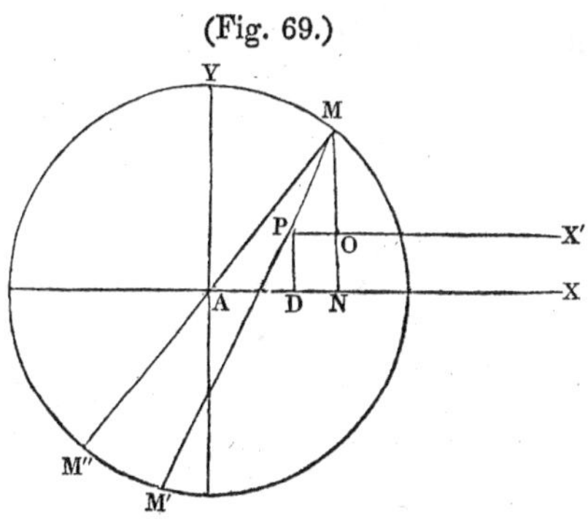

(Fig. 69.)

Put the radius vector PM$=r'$, the variable angle MPX′$=\omega$, AD and PD the co-ordinates of the pole$=m$ and n, and AN and MN the co-ordinates of M$=x$ and y.

Then MN=MO+ON=MO+PD=(169*b*) $r' \sin. \omega + n = y$.

And AN=DN+AD= PO+AD=(169*a*) $r' \cos. \omega + m = x$.

Squaring these values of x and y, substituting them into the equation of the circle referred to its centre, which is (173) $x^2+y^2-r^2=0$, and recollecting that $\sin.^2 \omega + \cos.^2 \omega = 1$, we obtain the equation

$$r'^2+2\,(m \cos. \omega + n \sin. \omega)\, r' + m^2+n^2-r^2=0,$$

which solved for r' gives

$$r' = -m \cos. \omega - n \sin. \omega \pm \sqrt{r^2-m^2-n^2+(m \cos. \omega + n \sin. \omega)^2},$$

which is the equation required, the two values of r' representing PM and PM′.

Strictly, however, it is only the positive value of r that we are to take into account, for PM′ is not truly the radius vector, but rather a continuation of it backward till it meets the curve in another point. The same will be true in all future cases when the value of the radius vector is negative.

(177) *Schol.* If the pole be placed on the line AX, it is evident that all the terms of the foregoing equation which contain n will

disappear; if it be situated on AY, all that contain m will disappear; and if at the origin A, both m and n will disappear, which will reduce the equation in the latter case to

$$r' = \pm r,$$

the polar equation when the pole is at the centre.

(178) Prop. V. Theorem.

The equation of an ellipse referred to its axes is $a^2y^2+b^2x^2-a^2b^2=0$; in which a and b are the semi-axes.

(Fig. 70.)

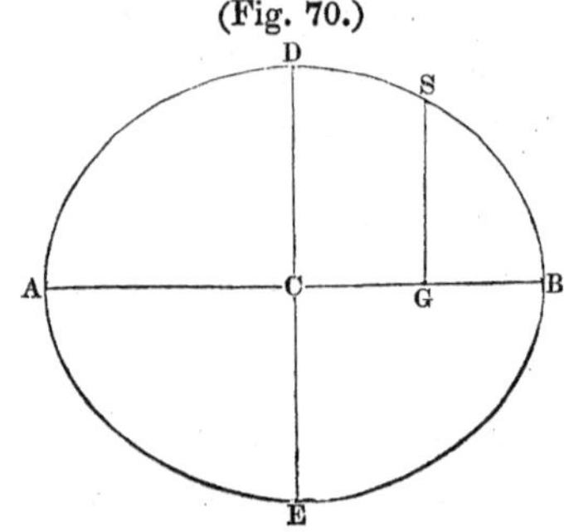

Let EBDA be the ellipse, AB and DE its axes, and CG and GS the co-ordinates of any point S.

Put AC$=a$.
" CD$=b$.
" CG$=x$.
" SG$=y$.

By (24) $SG^2 : AG.GB :: CD^2 : AC^2$.

But $AG=AC+CG$, and $GB=AC-CG$,

and consequently, $AG.GB=(AC+CG)(AC-CG)=$[a]AC^2-CG^2.

Therefore $SG^2 : AC^2-CG^2 :: CD^2 : AC^2$.

That is, $y^2 : a^2-x^2 :: b^2 : a^2$.

Converting this proportion into an equation, we have

$$a^2y^2=a^2b^2-b^2x^2,$$

which is the equation of an ellipse, and can readily be reduced to either of the following forms, viz.:

(178a) $$y^2=\frac{b^2}{a^2}(a^2-x^2), \quad \text{or} \quad a^2y^2+b^2x^2-a^2b^2=0.$$

---

[a] Leg. 4. 10. Euc. 2. 5, Cor.

(178*b*) *Schol.* By a similar process we can obtain from the property discussed in (54), the equation of an ellipse referred to any two conjugate diameters, viz.:

$$a'^2y^2+b'^2x^2-a'^2b'^2=0,$$

in which a' and b' represent the semi-conjugate diameters.

(179) *Cor.* If $a=b$, the ellipse becomes a circle; for then the last equation will reduce to $y^2+x^2-a^2=0$, which is (173) the equation of a circle.

(180) PROP. VI. PROBLEM.

To find the equation of a tangent to an ellipse.

Let MT be a line touching the ellipse at M, it is required to find its equation.

(Fig. 71.)

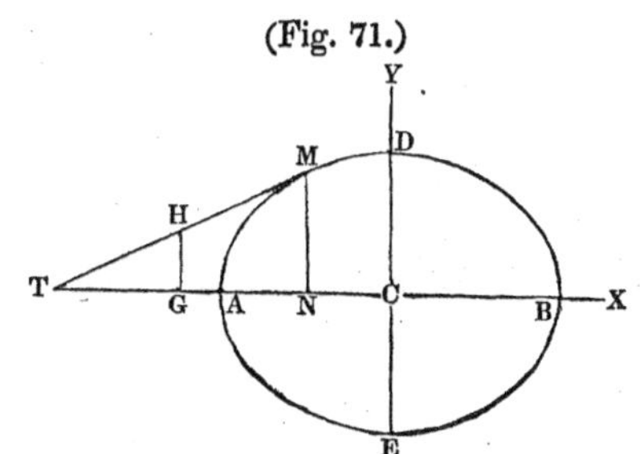

Since it passes through the point M its equation must (156) be of the form $y'-y=a'(x'-x)$, in which x' and y' are the co-ordinates of the point M, x and y those of any other point H in the line MT, and a' the tabular tangent of the angle MTX.

Put $AC=a$, and $CD=b$.

It was shown (24) that

$$CD^2 : AC^2 :: MN^2 : AN.NB=(39b)\ CN.NT.$$

$$\text{Hence} \quad NT=\frac{AC^2.MN^2}{CD^2.CN}=\frac{a^2y'^2}{b^2x'}.$$

Now, by trigonometry, $NT : MN :: R : \tan. MTX.$

$$\text{That is,} \quad \frac{a^2y'^2}{b^2x'} : y' :: 1 : a';$$

which reduced, gives $a'=\dfrac{b^2x'}{a^2y'}$.

It may be shown in the same manner as in the circle (175), that of the three quantities a', x', and y' in this equation, either one only or all three must in every case be negative, while b^2 and a^2 are always positive, and consequently that the foregoing equation will in all cases become

(180a) $$a' = -\frac{b^2x'}{a^2y'}.$$

Substituting this value of a' into the equation $y'-y=a'(x'-x)$, we obtain for the equation of the tangent MT,

(180b) $$y'-y=-\frac{b^2x'}{a^2y'}(x'-x).$$

This can be reduced to a more simple form by clearing it of fractions, and subtracting it from the equation of the ellipse, viz.: $a^2y'^2=a^2b^2-b^2x'^2$. It will then read, after transposition,

(180c) $$a^2yy'+b^2xx'-a^2b^2=0.$$

(181) Prop. VII. Problem.

To find the polar equation of an ellipse.

(Fig. 72.)

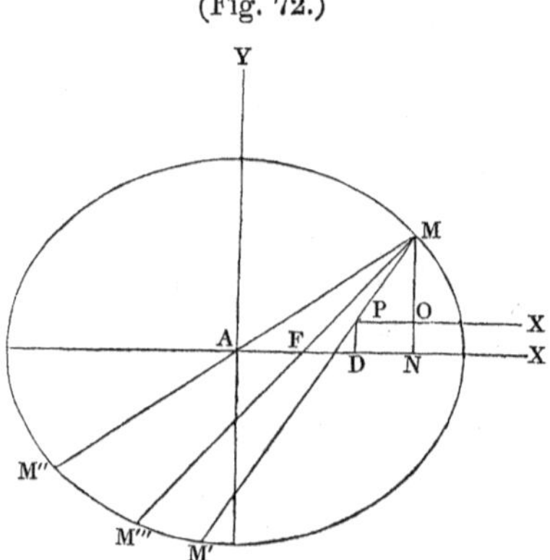

Let P be the pole, and PX′ parallel to AX the angular axis.

Put the radius vector PM $=r$, the variable angle MPX′ $=\omega$, AD and PD the co-ordinates of the pole $=m$ and n, and AN and NM the co-ordinates of the point $\text{M}=x$ and y.

Then, as in the circle (176), $x=r\cos.\omega+m$,

and $y=r\sin.\omega+n$.

Squaring these values of x and y, and substituting them into the equation of the ellipse (178), we obtain

$$(a^2 \sin.^2 \omega + b^2 \cos.^2 \omega)\, r^2 + 2\,(a^2 n \sin.\,\omega + b^2 m \cos.\,\omega)\, r + a^2 n^2 + b^2 m^2 - a^2 b^2 = 0,$$

which solved for r gives

(181a)
$$r = -\frac{a^2 n \sin.\,\omega + b^2 m \cos.\,\omega}{a^2 \sin.^2 \omega + b^2 \cos.^2 \omega} \pm \sqrt{\frac{a^2 b^2 - a^2 n^2 - b^2 m^2}{a^2 \sin.^2 \omega + b^2 \cos.^2 \omega} + \left(\frac{a^2 n \sin.\,\omega + b^2 m \cos.\,\omega}{a^2 \sin.^2 \omega + b^2 \cos.^2 \omega}\right)^2};$$

which is the equation required, the positive value of r representing PM, and the negative PM′.

(182) *Schol.* 1. If the pole be placed in the centre, the terms containing m and n disappear, and the equation reduces to

$$r = \pm \frac{ab}{\sqrt{a^2 \sin.^2 \omega + b^2 \cos.^2 \omega}};$$

making AM equal to AM″, as it evidently ought to be.

(183) *Schol.* 2. If the pole be placed at F, one of the foci, the polar equation may be obtained more easily by the following process.

In the equation of the ellipse referred to rectangular axes (178a), viz.: $y^2 = \frac{b^2}{a^2}(a^2 - x^2)$, substitute, in the place of b^2, its value (27) $a^2 - m^2$, and it will read

$$y^2 = \frac{a^2 - m^2}{a^2}(a^2 - x^2).$$

Now $\text{FM}^2 = \text{FN}^2 + \text{MN}^2 = (x - m)^2 + \frac{a^2 - m^2}{a^2}(a^2 - x^2) =$ (by reducing) $a^2 - 2mx + \frac{m^2 x^2}{a^2}$.

Extracting the roots of the first and last members, we have

$$\text{FM} = \pm\left(a - \frac{mx}{a}\right).$$

Putting r in the place of FM, and r cos. $\omega+m$ in the place of x, and reducing, we get for the two values of r,

(183a) $$r=-\frac{a^2-m^2}{a-m \text{ cos. } \omega}, \quad \text{and } r=\frac{a^2-m^2}{a+m \text{ cos. } \omega},$$

the values of FM‴ and FM.

These values of r can be expressed in a more simple form by dividing both numerator and denominator by a.

For (27) $\frac{a^2-m^2}{a}=\frac{b^2}{a}=$ (12) $\frac{1}{2}p$, p denoting the principal parameter.

Hence $$r=-\frac{\frac{1}{2}p}{1-\frac{m}{a}\text{ cos. } \omega}, \quad \text{or } r=\frac{\frac{1}{2}p}{1+\frac{m}{a}\text{ cos. } \omega}.$$

If $\omega=90°$ these expressions reduce to

(183b) $$r=\pm\tfrac{1}{2}p;$$

which is obviously true, for then r becomes FG (Fig. 2) $=\frac{1}{2}$GH $=$ (24a) $\frac{1}{2}p$.

We may also simplify the expressions for r in (183a) in another way, by introducing a letter that shall express the ratio of the eccentricity of the ellipse to the semi-transverse axis.

Put this ratio $=e=$ (14) $\frac{m}{a}$.

Hence $m=ae$, and $m^2=a^2e^2$.

Substituting these values in the place of m and m^2, and dividing both numerator and denominator by a, the expressions become

(183c) $$r=-\frac{a\,(1-e^2)}{1-e \text{ cos } \omega}, \quad \text{and } r=\frac{a\,(1-e^2)}{1+e \text{ cos. } \omega},$$

forms often met with.

(184) PROP. VIII. THEOREM.

The equation of a parabola referred to its vertex is $y^2=px$; in which p represents the principal parameter.

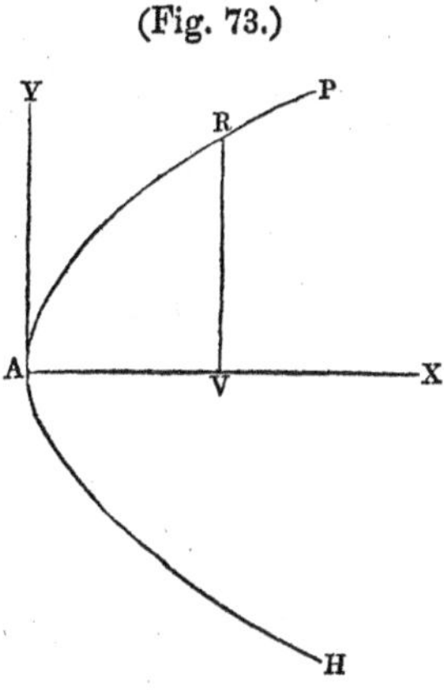

(Fig. 73.)

Let PAH be the parabola, A its vertex, and AV and RV the co-ordinates of any point R.

Put $AV=x$, and $RV=y$.

By (12) we have

$$x : y :: y : p.$$

Hence $y^2=px$.

(184*a*) *Schol.* By a similar process we can obtain the equation of a parabola referred to the vertex of any diameter, viz.:

$$y'^2=p'x';$$

in which p' represents the parameter of that diameter. (76*a*).

(185) PROP. IX. PROBLEM.

To find the equation of a tangent to a parabola.

Let MT be a line touching the parabola at M, it is required to find its equation.

Since it passes through the point M, its equation must (156) be of the form

(185*a*) $y'-y=a(x'-x)$;

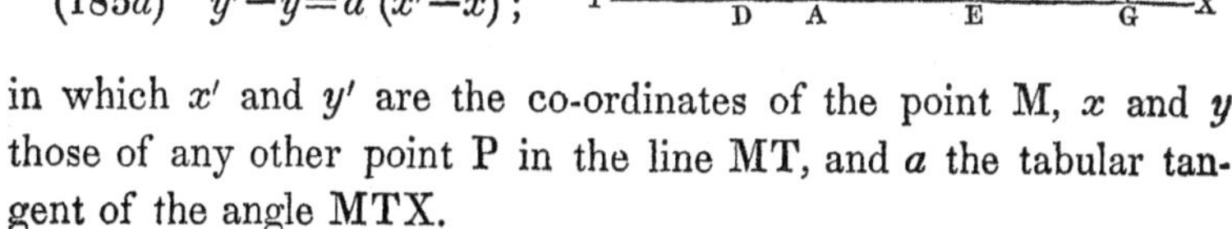

in which x' and y' are the co-ordinates of the point M, x and y those of any other point P in the line MT, and a the tabular tangent of the angle MTX.

By trigonometry, ET : ME :: R : tan. MTX.

That is (65), $2x' : y' :: 1 : a$;

which solved for a, gives

$$a=\frac{y'}{2x'}=\frac{y'^2}{2x'y'}=(184)\ \frac{px'}{2x'y'}=\frac{p}{2y'}.$$

Substituting the last value of a into (185a), we obtain for the equation of the tangent MT,

(185b) $$y'-y=\frac{p}{2y'}(x'-x);$$

or, by reducing, and substituting px in the place of y'^2,

(185c) $$yy'=\tfrac{1}{2}p\ (x'+x).$$

(186) Prop. X. Problem.

To find the polar equation of a parabola.

(Fig. 75.)

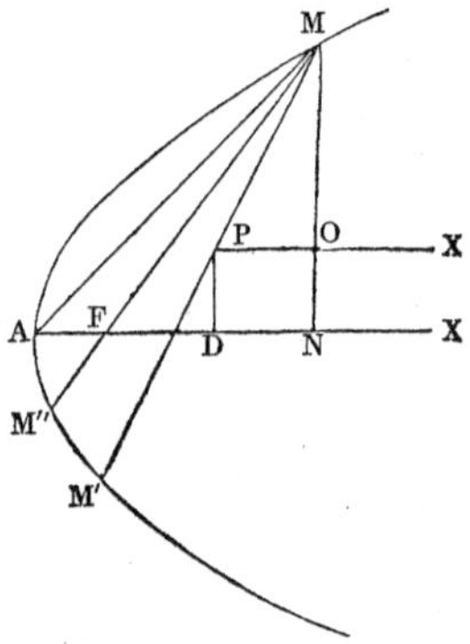

Let P be the pole, and PX′ parallel to AX, the angular axis.

Put the radius vector PM$=r$; the variable angle MPX′$=\omega$; AD and PD, the co-ordinates of the pole, $=m$ and n; and AN and MN, the co-ordinates of the point M, $=x$ and y.

Then, as in the circle (176),

$x=r\cos.\omega+m$, and $y=r\sin.\omega+n$.

Substituting these values of x and y into the equation of the parabola (184), and transposing, we obtain

$$r^2\sin.^2\omega+2\ (n\sin.\omega-\tfrac{1}{2}p\cos.\omega)\ r+n^2-pm=0;$$

which solved for r, gives

(186a) $$r=-\frac{n\sin.\omega-\frac{1}{2}p\cos.\omega}{\sin.^2\omega}\pm$$

$$\sqrt{\frac{pm-n^2}{\sin.^2\omega}+\left(\frac{n\sin.\omega-\frac{1}{2}p\cos.\omega}{\sin.^2\omega}\right)^2};$$

which is the equation required, the two values of r representing PM and PM′.

(187) *Schol.* 1. If the pole be placed at the vertex A, the terms that contain m and n disappear, and the equation reduces to

$$r=0, \quad \text{and} \quad r=\frac{p \cos. \omega}{\sin.^2 \omega},$$

the latter of which represents AM.

(188) *Schol.* 2. If the pole be placed at the focus F, $n=0$, and (59) $m=\frac{1}{4}p$.

Substituting this value of m into (186a), rejecting the terms that contain n, and recollecting that $\sin.^2 \omega+\cos.^2 \omega=1$, we obtain

$$r=\tfrac{1}{2}p\,\frac{\cos. \omega \pm 1}{\sin.^2 \omega};$$

the positive value being FM, and the negative FM″.

If $\omega=90°$ the last equation reduces to $r=\pm\frac{1}{2}p$.

(189) Prop. XI. Theorem.

The equation of an hyperbola referred to its axes is

$$a^2y^2-b^2x^2+a^2b^2=0;$$

in which a and b are the semi-axes.

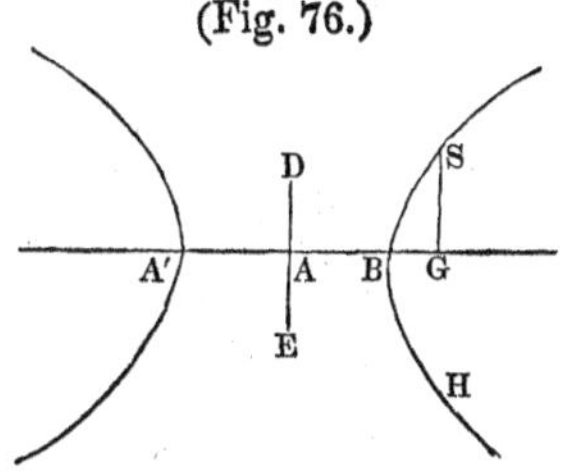

(Fig. 76.)

Let SBH be the hyperbola, A′B and DE its axes, and AG and SG the co-ordinates of any point S.

Put AB$=a$, AD$=b$, AG$=x$, and SG$=y$.

By (86)

$$SG^2 : A'G.GB :: AD^2 : AB^2.$$

But A′G$=$AG$+$AB, and GB$=$AG$-$AB; and consequently, A′G.GB$=$(AG$+$AB) (AG$-$AB)$=$[a]AG$^2-$AB^2.

---

[a] Leg. 4. 10. Euc. 2. 5, Cor.

Therefore SG^2 : AG^2—AB^2 :: AD^2 : AB^2.

That is, $y^2 : x^2-a^2 :: b^2 : a^2$;

which can be readily reduced to either of the following forms:

(189*a*) $$y^2=\frac{b^2}{a^2}(x^2-a^2), \text{ or } a^2y^2-b^2x^2+a^2b^2=0.$$

(189*b*) *Schol.* 1. The equation of the conjugate hyperbola is evidently

$$b^2x^2-a^2y^2+a^2b^2=0;$$

by merely interchanging the axes and co-ordinates.

(189*c*) *Schol.* 2. By a process similar to that employed in this proposition, we can obtain from (117) the equation of an hyperbola referred to any two conjugate diameters, viz.:

$$a'^2y^2-b'^2x^2+a'^2b'^2=0.$$

in which a' and b' represent the semi-diameters.

(190) Prop. XII. Problem.

To find the equation of a tangent to an hyperbola.

By a process analogous to that employed for the equation of a tangent to an ellipse (180), and which will be readily supplied by the student, we obtain the equation

$$y'-y=\frac{b^2x'}{a^2y'}(x'-x);$$

which may also, in the same manner as in (180*c*), be reduced to the form

$$a^2yy'-b^2xx'+a^2b^2=0.$$

(Fig. 77.)

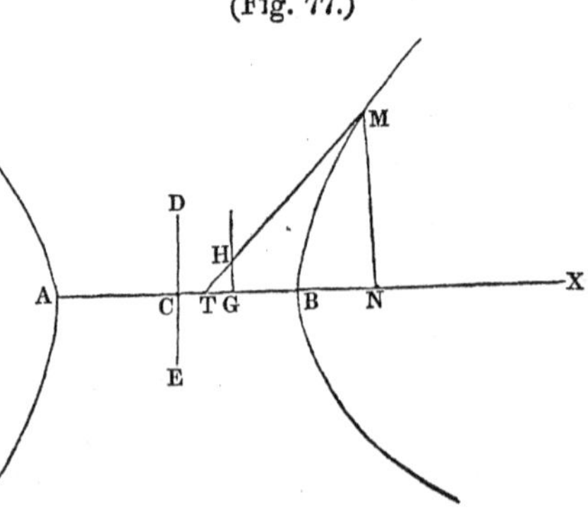

(191) PROP XIII. PROBLEM.

To find the polar equation of an hyperbola.

(Fig. 78.)

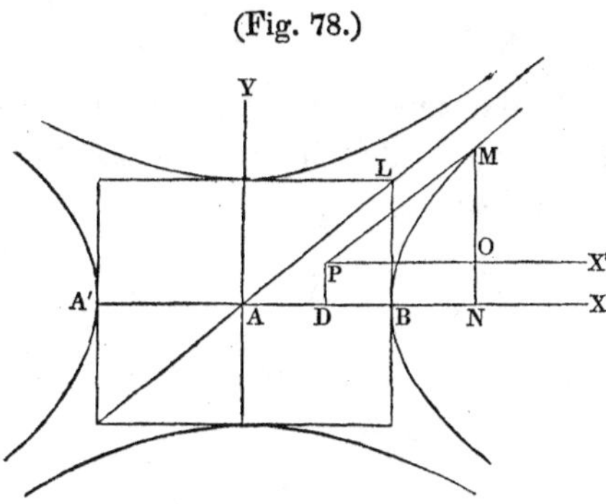

By the same process as in the ellipse (181), we obtain the equation

$$(a^2 \sin.^2 \omega - b^2 \cos.^2 \omega)\, r^2 + 2\,(a^2 n \sin.\,\omega - b^2 m \cos.\,\omega)\, r + a^2 n^2 - b^2 m^2 + a^2 b^2 = 0;$$

which solved for r gives

(191a)
$$r = -\frac{a^2 n \sin.\,\omega - b^2 m \cos.\,\omega}{a^2 \sin.^2 \omega - b^2 \cos.^2 \omega} \pm \sqrt{\frac{-a^2 b^2 - a^2 n^2 + b^2 m^2}{a^2 \sin.^2 \omega - b^2 \cos.^2 \omega} + \left(\frac{a^2 n \sin.\,\omega - b^2 m \cos.\,\omega}{a^2 \sin.^2 \omega - b^2 \cos.^2 \omega}\right)^2}.$$

(191b) *Schol.* 1. In the same manner as in the ellipse (182), we find that if the pole be placed in the centre A, the expressions be come

$$r = \pm \frac{ab}{\sqrt{-a^2 \sin.^2 \omega + b^2 \cos.^2 \omega}};$$

which obviously becomes impossible when $a \sin.\,\omega > b \cos.\,\omega$, showing that in that case the radius vector will not meet the curve. If $a \sin.\,\omega = b \cos.\,\omega$; that is, if $\frac{b}{a} = \frac{\sin.\,\omega}{\cos.\,\omega} =$ (by trigonometry) $\tan.\,\omega$, the denominator becomes o, which renders the value of r infinite.

When the radius vector is thus situated it is called an *assymptote;* and it is evident from the expression $\frac{b}{a}=\tan.\ \omega$, that it lies in the direction of the diagonal of a parallelogram described on the two axes; as AL.

(191*c*) *Schol.* 2. In the same manner, as in the ellipse (183), and by reference to (89*a*), we find that if the pole be placed at one of the foci, the expressions become

$$r=-\frac{a^2-m^2}{a-m\cos.\ \omega},\quad \text{and}\quad r=\frac{a^2-m^2}{a+m\cos.\ \omega};^{[a]}$$

which may be reduced to a more simple form by dividing both numerator and denominator by a.

$$\text{For}\quad \frac{a^2-m^2}{a}=(89a)\frac{-b^2}{a}=(12)-\tfrac{1}{2}p.$$

$$\text{Hence}\quad r=\frac{\frac{1}{2}p}{1-\frac{m}{a}\cos.\ \omega},\quad \text{and}\quad r=\frac{-\frac{1}{2}p}{1+\frac{m}{a}\cos.\ \omega}.$$

We may also, as in the ellipse (183*c*), simplify the expressions by introducing a letter (e) to represent the ratio of the eccentricity to the semi-transverse axis, which will reduce them to the form

$$(191d)\qquad r=-\frac{a\,(1-e^2)}{1-e\cos.\ \omega},\quad \text{and}\quad r=\frac{a\,(1-e^2)}{1+e\cos.\ \omega}.$$

(192) PROP. XIV. LEMMA.

Every equation between two variables of the form $cy^2+dx^2+e=0$ *is the equation of a conic section.*[b]

The three constants, c, d, and e, that enter into this equation, represent any known quantities whatever, whether positive or negative, and one of them must evidently have the contrary sign from the

---

[a] See Appendix, Note C.

[b] Including in this term the circle and straight line. See note on page 9.

other two, or the values of x and y will be imaginary. Hence the equation will reduce to one of the three forms

$$(192a)\quad \begin{cases} cy^2+dx^2-e=o. \\ cy^2-dx^2+e=o. \\ dx^2-cy^2+e=o. \end{cases}$$

Multiply each by e and divide by cd, and they become

$$(192b)\quad \begin{cases} \frac{e}{d}y^2+\frac{e}{c}x^2-\frac{e^2}{cd}=0. \\ \frac{e}{d}y^2-\frac{e}{c}x^2+\frac{e^2}{cd}=0. \\ \frac{e}{c}x^2-\frac{e}{d}y^2+\frac{e^2}{cd}=0. \end{cases}$$

Putting $\frac{e}{d}=a^2$, and $\frac{e}{c}=b^2$, and substituting, they become

$$(192c)\quad \begin{cases} a^2y^2+b^2x^2-a^2b^2=0, \text{ the ellipse (178}b\text{).} \\ a^2y^2-b^2x^2+a^2b^2=0, \text{ the hyperbola (189}a\text{).} \\ b^2x^2-a^2y^2+a^2b^2=0, \text{ the conjugate hyperbola (189}b\text{).} \end{cases}$$

If in the original equation $e=o$, c and d must have contrary signs, and the equation will reduce to $y=\left(\frac{d}{c}\right)^{\frac{1}{2}}x$, which is (154) the equation of a straight line passing through the origin. Or, if in the first of (192a) $c=d$, the equation will reduce to $y^2+x^2=\frac{e}{c}$, the equation of a circle referred to its centre (173).

The conditions of the proposition forbid that either c or d should become zero, for then would one of the variables disappear.

(193) PROP. XV. THEOREM.

Every equation of the second degree between two variables is the equation of a conic section.

Every such equation can be reduced, by transposition, multiplication, and division, to the form

$$(193a)\qquad Ay^2+Bxy+Cx^2+Dy+Ex+F=0;$$

in which the co-efficients, A, B, C, &c., represent any known quantities, whether positive or negative. This equation solved for y gives

(193*b*)
$$y=-\frac{1}{2A}(Bx+D)\pm$$
$$\frac{1}{2A}\sqrt{(B^2-4AC)x^2+2(BD-2AE)x+(D^2-4AF)};$$

in which we will, for the sake of simplicity, put

$$B^2-4AC=p,\quad BD-2AE=q,\quad \text{and } D^2-4AF=s;$$

which will reduce the equation to the form

(193*c*)
$$y=-\frac{1}{2A}(Bx+D)\pm\frac{1}{2A}\sqrt{px^2+2qx+s}.$$

This value of y consists of two parts, and if the first part were taken by itself, we should have the equation of a straight line, viz.:

(193*d*)
$$y=-\frac{1}{2A}(Bx+D)=-\frac{B}{2A}x-\frac{D}{2A};$$

in which $-\frac{B}{2A}$ corresponds to a in the general form (153), and $-\frac{D}{2A}$ to b.

The second part being either positive or negative, shows that each ordinate meets the curve in two points, one as far above the line of which (193*d*) is the equation, as the other is below it, and consequently that this line bisects the curve, or is its diameter.

At the points where this line intersects the curve, if at all, the second part of the value of y in (193*c*) must be zero, and we therefore have for these points the equation

(193*d*2)
$$px^2+2qx+s=0;$$

which solved for x gives

(193*e*)
$$x=-\frac{q}{p}\pm\frac{1}{p}\sqrt{-ps+q^2}.$$

Hence for the centre, or point midway between the intersections, we have

(193*f*) $$x = -\frac{q}{p};$$

and substituting this value of x into (193*d*), we obtain

(193*g*) $$y = -\frac{1}{2A}\left(-B\frac{q}{p} + D\right).$$

Equations (193*f*) and (193*g*) make known the co-ordinates of the centre of the curve, if it have a centre.

The preceding discussion will be better understood by the aid of a diagram.

(Fig. 79.)

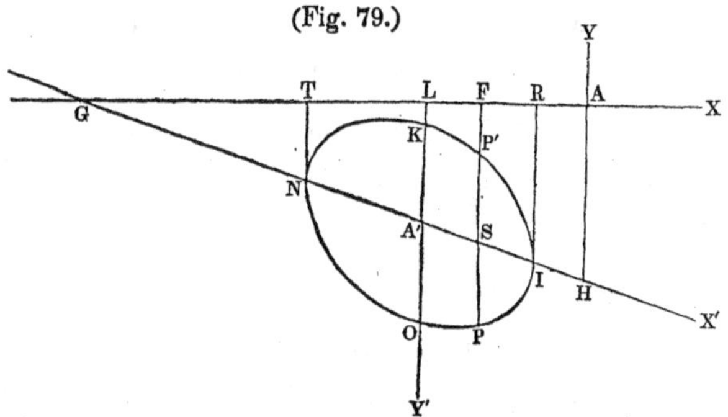

Let GX′ and IKNO be the straight line and curve, of which (193*d*) and (193*c*) are the equations, drawn according to (170). As $-\frac{B}{2A}$ represents the tangent of the angle which GX makes with the axis of abscissas, reckoned in the usual direction, we have

(193*h*) $$-\frac{B}{2A} = \tan.\ AGX' = {}^{a}\frac{\sin.\ AGX'}{\cos.\ AGX'}.$$

Since the members of this equation have opposite signs, it follows that the sine and cosine of AGX′ have the same sign when the signs of A and B are unlike, and contrary signs when those of A and B are alike. In constructing the diagram the signs of A

---

[a] Davies' Leg. Trig. Sec. XVII.

and B were supposed to be alike; otherwise the line GX′ would have been situated in the first and third quadrants, instead of the second and fourth, as it now is.

It is also to be noticed that the line AH, represented in (193*d*) by the fraction $-\frac{D}{2A}$, falls below the axis of abscissas when the signs of A and D are alike, but above it when they are unlike.

The values of y in (193*c*) represent any ordinate to the curve, as PF, meeting the curve at two points, P and P′; the first part representing FS, and the second PS or P′S.

The values of x in (193*e*) represent AR and AT, the abscissas of the points I and N; and the values of x and y in (193*f*) and (193*g*), the co-ordinates of the point A′, midway between I and N.

We will now, by a transformation of co-ordinates (162), refer the curve to the lines GA′X′ and LA′Y′ as new axes, transferring the origin of co-ordinates from A to A′.

The equations for transformation (162*a* and 162*b*) are

$$(193l) \quad \begin{cases} x' \cos. a + y' \cos. a' + m = x, \\ x' \sin. a + y' \sin. a' + n = y; \end{cases}$$

in which, when applied to the present case, a represents the angle AGX′, a' the angle ALY′, both estimated in the usual direction, m and n the co-ordinates of A′, viz.:

$$m = -\frac{q}{p} = \text{AL}, \quad \text{and } n = -\frac{1}{2A}(-B\frac{q}{p} + D) = \text{A'L};$$

and x' and y' the co-ordinates of any point in the curve referred to the new axes. Since, as here drawn,[a] the angle AGX′ terminates in the fourth quadrant, its sine is negative and its cosine positive.[b]

We have also $\cos. a' = \cos. 270° = 0$, and $\sin. a' = \sin. 270° = -1$.

---

[a] The general process is the same for any other position of the line GA′X, regard being had to the algebraic signs.

[b] Davies' Leg. Trig. Sec. XII.

Hence the equations in (193*l*) become

$$(193m)\quad \begin{cases} x' \cos. a - \dfrac{q}{p} = x, \\ -x' \sin. a - y' - \dfrac{1}{2A}\left(-B\dfrac{q}{p} + D\right) = y; \end{cases}$$

Substituting these values of x and y into (193*c*), and cancelling equal terms, we obtain

$$(193n)\quad -x' \sin. a + y' = \frac{B}{2A} x' \cos. a \pm \frac{1}{2A}\sqrt{px'^2 \cos.^2 a - \frac{q^2}{p} + s}.$$

But (193*h*) shows that $\dfrac{B}{2A} \cos. a = -\sin. a$; consequently (193*n*) reduces to

$$(193o)\quad y' = \pm \frac{1}{2A}\sqrt{px'^2 \cos.^2 a - \frac{q^2}{p} + s}.$$

Squaring, clearing of fractions, and transposing, we obtain

$$(193p)\quad 4A^2 y'^2 - px'^2 \cos.^2 a + \frac{q^2}{p} - s = 0.$$

Or, if we put $4A^2 = c$, $-p \cos.^2 a = d$, and $\dfrac{q^2}{p} - s = e$, the equation will read

$$cy'^2 + dx'^2 + e = 0;$$

which, by (192), is the equation of the conic section.

(193*r*) *Schol.* There is a single case not provided for in the foregoing demonstration. If $B^2 = 4AC$, the value of p becomes zero, and consequently that of x in (193*f*) infinite, showing that when that relation exists the curve cannot have a centre, and consequently that the new system of co-ordinates cannot be referred to it. In that case (193*d*2) will become

$$2qx + s = 0, \quad \text{or } x = -\frac{s}{2q};$$

which will reduce (193*d*) to the form

$$y = \frac{Bs}{4Aq} - \frac{D}{2A}.$$

The last two equations make known the co-ordinates of the point where the line GX′, of which (193*d*) is the equation, and which remains unchanged, intersects the curve, as at N in the diagram. We may therefore refer the curve to the lines NX′ and NT as new axes, transferring the origin to N instead of A′.

The equations for transformation corresponding to those in (193*m*), will be

$$x' \cos. a - \frac{s}{2q} = x,$$

$$-x' \sin. a - y' + \frac{Bs}{4Aq} - \frac{D}{2A} = y.$$

By substituting these values of x and y into (193*c*), and taking the same steps we did to obtain (193*n*) and (193*o*), we get

$$y' = \pm \frac{1}{2A} \sqrt{2qx' \cos. a};$$

which squared, gives

$$y'^2 = \frac{q \cos. a}{2A^2} x' = p'x, \text{ putting } p' \text{ for } \frac{q \cos. a}{2A^2}.$$

Hence the curve is a parabola (184*a*).

(194) Prop. XVI. Problem.

The equation of the tangent to any curve being given, to find the equation of the normal.

Let CMB be the curve, MT the tangent, and MG the normal.

(Fig. 80.)

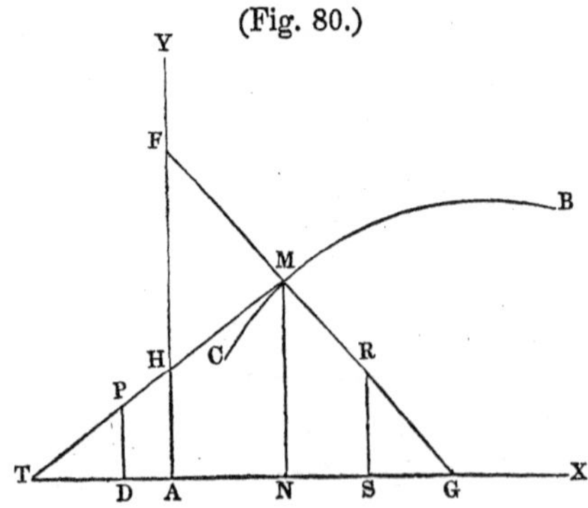

Since both these lines pass through the point M, their equations will (156) be of the form

$$(194a) \quad \begin{cases} y'-y=a\ (x'-x), & \text{the equation of MT,} \\ y'-y=a'\ (x'-x), & \text{the equation of MG;} \end{cases}$$

in which x' and y' are the co-ordinates of the point M, x and y those of any other point in the line MT or MG, a the tangent of the angle MTX, and a' the tangent of the angle MGX. Moreover, since GM is perpendicular to MT, we have, by (161),

$$(194b) \qquad 1+aa'=0.$$

By substituting into (194b) the value of a taken from the given equation of MT, we may obtain the value of a', and this value again substituted into the above equation of MG, gives us the equation required.

To illustrate, we will suppose the curve to be a circle referred to its centre. The equation of its tangent (175b) is

$$y'-y=-\frac{x'}{y'}(x'-x);$$

in which $-\frac{x'}{y'}$ corresponds to a in the general formula (194a).

Substituting this value of a into the formula $1+aa'=o$, we obtain

$$1-\frac{x'}{y'}a'=0;$$

which solved for a' gives

$$a'=\frac{y}{x'}.$$

Substituting this value of a' into the equation of MG in (194a), we have

$$y'-y=\frac{y'}{x'}(x'-x);$$

which is the equation of the normal.

Example. Find the equation of the normal to the ellipse, parabola, and hyperbola.

Ans. $y'-y=\frac{a^2y'}{b^2x'}(x'-x)$, for the ellipse.

$y'-y=-\frac{2y}{p}(x'-x)$, for the parabola

$y'-y=-\frac{a^2y'}{b^2x'}(x'-x)$, for the hyperbola.

(195) Prop. XVII. Problem.

The equation of a tangent or normal to a curve being given, to find the points where it intersects the axes of reference.

Let CMB be the curve, AX and AY the axes, MT the tangent, and MG the normal at the given point M; it is required to find the distances AT, AH, AG, and AF.

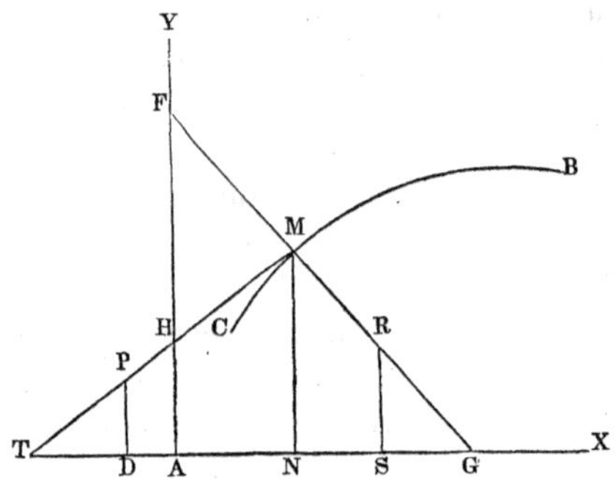

The equation of the tangent is (156) of the form

$$y'-y=a(x'-x);$$

in which x' and y' are the co-ordinates of the given point M, and x and y those of any point P in the tangent MT. As this equation is true wherever the point P be taken, we may suppose P to move towards T till the ordinate PD becomes o, and x=AT. The equation will then read

$$y'-o=a(x'-x);$$

in which all the quantities except x are known, and consequently its value may be found, which gives the length of AT.

In like manner, by moving P the other way till it coincides with H, we shall have $x=o$, and y=AH, which reduces the equation to

$$y'-y=a(x'-o).$$

This solved for y makes known the length of AH.

The lengths of AG and AF are found in the same way by using the equation of the normal (194), and supposing the point R to move first to G and then to F.

Ex. 1. Let us suppose the curve to be a parabola whose principal parameter is 9 inches, the abscissa AE $=4$ inches, and the ordinate ME 6 inches.[a] It is required to find the lengths of AT, AH, AG, and AF.

By the equation of the tangent (185), and normal (194), we have

For the point T, $y'-o=\frac{p}{2y'}(x'-x)$; that is, $6-o=\frac{9}{12}(4-x)$.

Hence $x=-4=$ AT.

For the point H, $y'-y=\frac{p}{2y'}(x'-o)$; that is, $6-y=\frac{9}{12}(4-o)$.

Hence $y=3=$ AH.

For the point G, $y'-o=-\frac{2y'}{p}(x'-x)$; that is, $6-o=-\frac{12}{9}(4-x)$.

Hence $x=8\frac{1}{2}=$ AG.

For the point F, $y'-y=-\frac{2y'}{p}(x'-o)$; that is, $6-y=-\frac{12}{9}(4-o)$.

Hence $y=11\frac{1}{3}=$ AF.

Ex. 2. Let the curve be a circle referred to its centre (173), the radius being 10 inches, the abscissa -6 inches, and the ordinate 8 inches. Required as above.

Ans. AT$=-16\frac{2}{3}$ inches, AH$=12\frac{1}{2}$, AG$=0$, and AF$=0$.

We discover from this example that any normal to a circle passes through the centre.

Ex. 3. Let the curve be an ellipse referred to its axes (178), the transverse axis being 10 inches and the conjugate 8 inches,

---

[a] In this problem and several which follow, it would be sufficient to give the value of but one of the co-ordinates, x or y, as the other could be found from it by means of the equation of the curve; but it would render the solutions more complex when the equation is above the first degree, since more than one value could be found that would satisfy the conditions of the problem.

the abscissa 3 inches and the ordinate $3\frac{1}{5}$ inches. Required as above.

Ans. AT$=8\frac{1}{3}$ inches, AH$=5$, AG$=1\frac{2}{25}$, and AF$=-1\frac{4}{5}$.

(195*a*) *Cor.* We are enabled by this proposition to determine the area of the triangle AHT or AGF, which the tangent or normal forms with the axes of reference.

(196) Prop. XVIII. Problem.

Two curves whose equations are known intersect one another, to determine the point of intersection.

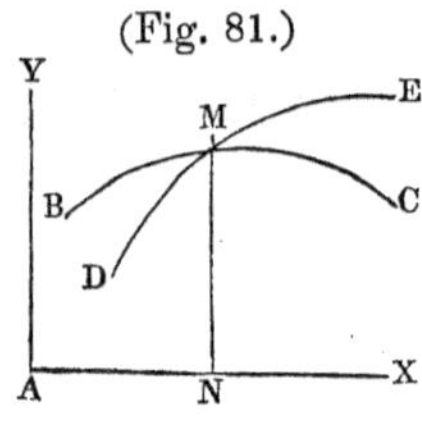

(Fig. 81.)

Let BMC and DME be the two curves, and M the point of intersection.

At the point M the co-ordinates AN and MN are common to both curves, so that we have but two unknown quantities, x and y; and we have two equations, viz.: the equations of the two curves, by which to find these values. Solve these equations for x and y, and we have the co-ordinates of the point required.

If the curves are not referred to the same axes, it will be necessary as a preliminary step to reduce them to the same, in the manner detailed in (162 to 168).

Ex. 1. Let the curves be $y^2=px$, a parabola (184);
and $x^2+y^2=r^2$, a circle (173);

in which $p=10$, and $r=12$.

Solving these equations for x and y, we obtain

$$x=-\tfrac{1}{2}p\pm\sqrt{r^2+\tfrac{1}{4}p^2}=8,$$

$$y=\sqrt{-\tfrac{1}{2}p^2\pm p\sqrt{r^2+\tfrac{1}{4}p^2}}=\pm\sqrt{80}.$$

Ex. 2. Let the curves be

$$a^2y^2+b^2x^2-a^2b^2=0, \text{ an ellipse;}$$

$$\text{and } a'^2y^2-b'^2x^2+a'^2b'^2=0, \text{ an hyperbola;}$$

in which $a=10$, $a'=8$, $b=8$, and $b'=6$.

Ans. $x=\pm9,12$ nearly; and $y=\pm3,3$ nearly.

(197) Prop. XIX. Problem.

Determine the angle formed by the intersection of two curves, the equations of whose tangents are given.

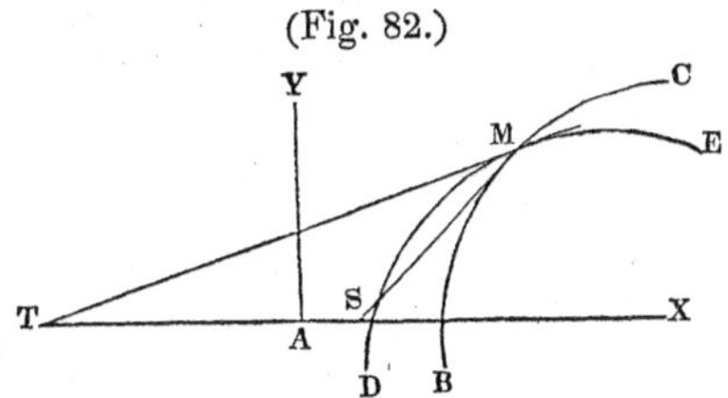

(Fig. 82.)

Let BMC and DME be the two curves, and MS and MT tangents to them at the point of intersection M.

The equations of MS and MT make known the tangents of the angles MSX and MTX, from which we can obtain, by (160), the angle SMT, which is the same as that made by the curves at the point of intersection.

Ex. 1. Let DME be a parabola, and BMC an hyperbola.

Then (190) the equation of MS is $y'-y=\frac{b^2x'}{a^2y'}(x'-x)$.

And (185*b*) the equation of MT is $y'-y=\frac{p}{2y'}(x'-x)$.

Hence tan. $\text{MSX}=\frac{b^2x'}{a^2y'}$, and tan. $\text{MTX}=\frac{p}{2y'}$.

Substituting these values in the place of a and a' in the formula (160), reducing, and for y'^2 substituting its equal (184) px', we get

$$\text{tan. SMT}=\frac{2b^2x'-a^2p}{(2a^2+b^2)\,y'}.$$

All the quantities in this expression being supposed to be known, we have the angle SMT.

Ex. 2 and 3. Find the angles at which the curves in Prop. XVIII. intersect.

Ans. The parabola and circle at an angle of 71° 0′ 58″.
The ellipse and hyperbola at an angle of 62° 14′ 4″.

(198) Prop. XX. Problem.

Given the equation of a curve and of its tangent, to find the point on the curve at which, if a tangent be drawn, it will make a given angle with the axis of abscissas.

(Fig. 83.)

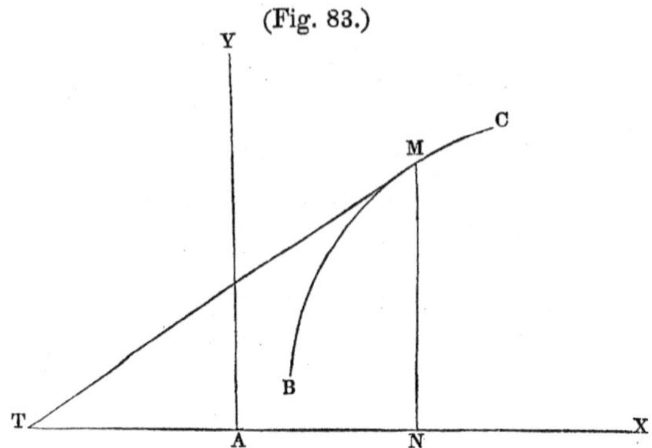

Let BMC be the given curve, and MTX the given angle.

Put tan. MTX$=m$.

The equation of MT makes known the value of MTX in terms of x', y', and constants, which being put equal to m gives us one equation, and this, together with the equation of the curve, will enable us to find the values of the co-ordinates x' and y'.

Ex. 1. Let the equation of the curve be $x'^2+y'^2-r^2=0$, a circle, the equation of whose tangent (175*b*) is $y'-y=-\frac{x'}{y'}(x'-x)$.

Hence tan. MTX$=-\frac{x'}{y'}=m$.

Solving this and the equation of the curve for x' and y', we get

$$x'=\pm\frac{mr}{\sqrt{m^2+1}}, \quad \text{and} \quad y'=\pm\frac{r}{\sqrt{m^2+1}}.$$

Ex. 2. Let the curve be a parabola, whose parameter is 9 inches, and the given angle 20°.

Ans. $x'=16{,}984$, and $y'=12{,}364$.

(199) PROP. XXI. PROBLEM.

To find where the tangent to a curve at a given point will intersect another curve, the equation of the second curve, and that of the tangent to the first being given.

This is merely a particular case of Prop. XVIII., and is solved in the same manner, the two equations which determine the values of x and y being that of the tangent line and of the second curve.

Ex. 1. Let a tangent to a parabola intersect a curve whose equation is $xy=a$.

The equation of a tangent to a parabola (185*b*) is

$$y'-y=\frac{p}{2y'}(x'-x).$$

Solving these two equations for x and y, and substituting px' in the place of y'^2, we obtain

$$x=-\tfrac{1}{2}x'\pm\sqrt{\frac{2ay'}{p}+\tfrac{1}{4}x'^2},\quad\text{and}\quad y=\frac{a}{-\tfrac{1}{2}x'\pm\sqrt{\frac{2ay'}{p}+\tfrac{1}{4}x'^2}}.$$

If we give numerical values to the letters, as $a=10$ inches, $p=9$, $x'=4$, and $y'=6$, we have $x=+2{,}163$ inches nearly, or $-6{,}163$ nearly; and $y=+4{,}62$ nearly, or $-1{,}62$ nearly.

Ex. 2. Let a tangent to a circle whose radius is 10 inches meet a parabola whose parameter is 9, the co-ordinates of the point of tangency being $x'=8$, and $y'=6$, and the vertex of the parabola and also the origin of the co-ordinates being the centre of the circle.

Ans. $x=6{,}68$, or $23{,}36$; and $y=7{,}756$, or $-14{,}5$.

(200) PROP. XXII. PROBLEM.

To determine the distance from a given point to a tangent to a curve, the equation of the tangent being given.

Let P be the given point, and TM the given tangent to the curve BMC at the point M, it is required to find the distance PH perpendicular to TM; and as a preliminary step to find AE and EH, the co-ordinates of the point H; for if these be known, the distance PH can be determined by (159).

(Fig. 84.)

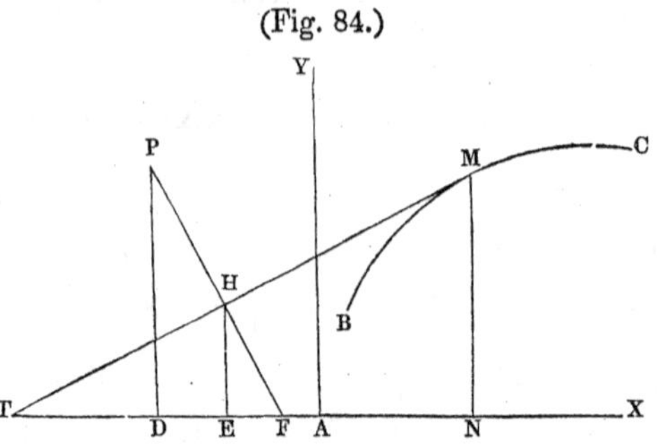

Produce PH till it meets the axis of abscissas in F.

Put $AN=x'$, $MN=y$, $AE=x$, $EH=y$, $AD=m$, $DP=n$, tan. $HTX=a$, and tan. $HFX=a'$.

Since THF is a right angle we have, by (161), the equation

$$1+aa'=0, \quad \text{or } a'=-\frac{1}{a};$$

which makes known the value of a', that of a being given in the equation of TM.

Since the line TM passes through the given point M, and PF through P, their equations will be

(200a) $\quad y'-y=a\ (x'-x)$, the equation of TM.

(200b) $\quad n-y=a'\ (m-x)$, the equation of PF.

These equations solved for x and y make known the co-ordinates of the point H, and we can then find the distance PH by the formula at (159), viz.: $PH=\sqrt{(m-x)^2+(n-y)^2}$.

Ex. 1. Let the curve be a circle referred to its centre, the equation of whose tangent is (175b)

(200c) $\quad y'-y=-\dfrac{x'}{y'}(x'-x)$, the equation of TM;

in which $-\dfrac{x'}{y'}$ corresponds to a in the general form.

Substituting this value of a into the formula $1+aa'=0$, and solving for a', we get

$$a'=\frac{y'}{x'}.$$

This value of a' substituted into the equation of PF (200*b*), gives

$$(200d) \qquad n-y=\frac{y'}{x'}(m-x).$$

Solving equations (200*c* and 200*d*) for x and y, and recollecting (173) that $x'^2+y'^2=r^2$, we obtain

$$x=x'+\frac{(my'-nx')\,y'}{r^2}, \quad \text{and} \quad y=y'+\frac{(nx'-my')\,x'}{r^2}.$$

If $r=10$, $x'=8$, $y'=6$, $m=12$, and $n=5$, we shall find

$$x=9{,}92, \quad y=3{,}44, \quad \text{and PH}=2{,}6.$$

Ex. 2. A comet moving from C towards B in the parabolic orbit CMB, whose parameter is 150 millions of miles, its vertex at A, and its transverse AX, arrives at the point M, where the ordinate MN is 100 millions of miles, and at that point flies off from its orbit in the direction of the tangent MT. The earth, at the time the comet passes it, is at P, where the ordinate PD is 7 millions of miles, and the abscissa AD $57\frac{1}{3}$ millions. How far does the comet pass from the earth?

Ans. It strikes it.

(201) *Promiscuous Examples.*

1. A circle whose radius is 10 touches externally an ellipse whose transverse axis is 10 and its conjugate axis 8; the abscissa of the point of contact referred to the axes of the ellipse is 3. Required the position of the centre of the circle.

Ans. The abscissa of the centre referred to the axes of the ellipse is 8,145; and the ordinate $\pm 11{,}78$.[a]

2. Find the point on a parabola, whose parameter is 9, at which if a tangent and normal be drawn, the triangles which they form

[a] The negative value of x and the corresponding values of y are the co-ordinates of the centre of an inscribed circle touching the ellipse at the same point.

with the axes of reference (195a) shall be to each other in the ratio 1 : 8.

Ans. $x=4,5$, and $y=\sqrt{40,5}$.

3. Find the point on a parabola, whose parameter is 9, at which if a tangent and normal be drawn, they will form with the transverse axis a triangle whose area is 100.

Ans. $x=8,92$, and $y=8,96$.

4. Find the point on an hyperbola whose transverse axis is 10, and conjugate axis 8, at which if a tangent and normal be drawn, the subtangent will be to the subnormal in the ratio 2 : 5.

Ans. $x=5,8$, and $y=2,35$.

5. Any number of curves intersect; it is required to find the distance between any two points of intersection.[a]

Let there be four curves whose equations are: 1st, $y=x-4$; 2d, $y^2=7x$; 3d, $x^2+y^2=100$; and 4th, $y\ (x-3)=8$.

Find the distance from the intersection of the first and fourth above the axis of abscissas, to the intersection of the second and third below the axis of abscissas. Also, find the distance from the intersection of the first and second, to the intersection of the second and third, both above the axis of abscissas.

Ans. The distances are 9,44 and 7,31.

6. Find whether the lines, whose lengths were found in the last example, intersect; and if so, where, and at what angle.[b]

Ans. They will intersect, if produced, at an angle of 72° 27′; and the abscissa of the point of intersection will be 5,91, and the ordinate 6,55.

---

[a] The co-ordinates of the points of intersection being found by (196), the distance may be found by (159).

[b] Each of the lines passes through two given points; hence (157) their equations are of the form $y'-y=\frac{y''-y'}{x''-x'}(x'-x)$, in which x', y', x'', and y'' are the co-ordinates found in the last example. Substituting the numerical values in each, we shall have two equations, by means of which the point of intersection may be found by (196), and the angle by (160), the values of the fraction $\frac{y''-y'}{x''-x'}$ in the two equations, corresponding to a and a' in the formula.

CHAPTER III.

OF LINES IN SPACE.

(202) If three planes of indefinite extent intersect each other, as in the figure, they will divide all space into eight parts; and as in a plane the position of a point is determined by drawing ordinates to two given axes lying in the plane, so the position of any point in space may be determined by drawing ordinates to these three *co-ordinate* planes, as they are usually styled. For the sake of simplicity, the three planes are usually taken at right angles to each other when practicable, as represented in the figure, and in our future discussions they are to be so regarded. The plane STVH is called the *horizontal* plane, and the other two the *vertical* planes. The lines X′AX, Y′AY, and Z′AZ, in which the planes intersect, are called *axes;* and the point A, in which the axes intersect, the *origin*. Ordinates in the several directions, AX, AY, and AZ, are denoted respectively by the letters x, y, and z; and those in the opposite directions by the same letters with the negative sign prefixed, in the same manner as in equations of lines in a plane. Unlike those equations, however, AZ is usually taken as the axis of abscissas, and the other two as axes of ordinates.

(Fig. 86.)

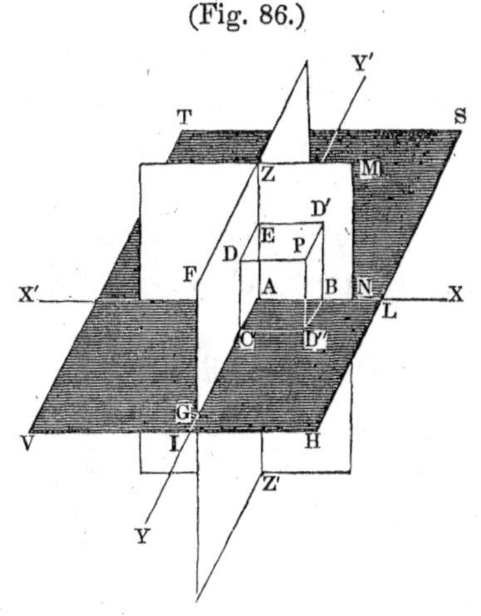

(203) Instead of giving the lengths of the ordinates themselves to determine the position of a point, we may give the measure of them on the axes to which they are parallel, in the same manner as in the plane. Thus the position of P may be determined by giving the lengths of PD, PD′, and PD″, or by giving the lengths of AB, AC, and AE.

(204) The points where the ordinates of a point in space meet the several co-ordinate planes are called *projections;* and, in like manner, the lines traced upon the co-ordinate planes by an indefinite number of ordinates let fall upon them from any line in space, whether straight or curved, are called the projections of that line. Thus D is the projection of the point P upon the plane AZFG, D′ its projection upon ANMZ, and D″ its projection upon AIHL. Also, the lines ED′ and CD″ are the projections of the line PD upon the two latter planes. Being perpendicular to the plane AZFG, it is projected upon it in a point at D; but if it were oblique it would be projected into a line upon this plane also. The plane in which both a line and its projection lie is called the *projecting* plane.

(205) *Def.* A *cylindrical surface* is one generated by a straight line moving parallel to itself, while its extremity describes a curve.

(206) *Def.* A *conical surface* is one generated by a straight line passing through a fixed point, while its extremity describes a curve. It obviously consists of two parts united only at the fixed point or *vertex.* The two parts are called *sheets* or *nappes.*

(207) The moving line in (205) and (206) is called the *generatrix,* and the curve the *directrix.*

(208) *Def.* A *conoid* is a solid generated by the revolution of either of the conic sections about one of its axes, and is either an *ellipsoid,* a *paraboloid,* or an *hyperboloid,* according as the genera trix is an ellipse, a parabola, or an hyperbola.

(209) Prop. I. Theorem.

The projections of a point or straight line, or either two of the co-ordinate planes, determine its position.[a]

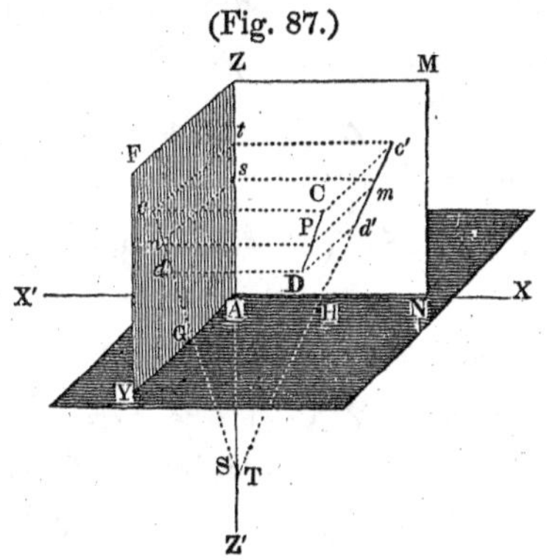

(Fig. 87.)

Firstly. Let C be the point, and c and c' its projections upon the vertical co-ordinate planes AZFY and ANMZ.

At the points c and c' draw perpendicular to their respective planes the lines cC and c'C. Since the point C must by (204) lie in both these perpendiculars, their mutual intersection must determine its position.

Secondly. Let CD be the straight line, and cd and $c'd'$ its projections upon the vertical planes.

Through cd and $c'd'$ draw the planes cCDd and c'CDd' perpendicular to the co-ordinate planes, and since the line CD must lie in both these planes, their mutual intersection must determine its position.

(209*a*) *Schol.* This proposition applies also to any plane curve; for such a curve must lie in each of two right cylindrical surfaces, the projections upon the co-ordinate planes being the directrices.

(210) Prop. II. Theorem.

The equations of a straight line in space are

$$x = az + \alpha, \quad \text{and} \quad y = bz + \beta;$$

in which x, y, and z represent the co-ordinates of any point in the line, a and b the tangents of the angles which the projections of the line upon the vertical planes make with the axis of abscissas, and α and β the parts of the axes of ordinates intercepted between these projections and the origin.

---

[a] If the line is parallel to one of the co-ordinate planes, one of the projections must be taken on that plane.

Let CD (Fig. 87) be the straight line, and cd and $c'd'$ its projections.

Also let P be any point in CD, and m and n its projections.

Produce $c'd'$ and cd, if necessary, till they meet the axis of abscissas in T and S, and draw the ordinates ms and ns.

Then $ms = Pn = x$, $ns = Pm = y$, and by (203) $As = z$.

Put tan. $c'TZ$ (the angle being estimated from Z towards the right) $= a$.

Put tan. cSZ (the angle being estimated from Z towards the left) $= b$.

Also put $AH = \alpha$, and $AG = \beta$.

Then, since Z′AZ is the axis of abscissas, AX the axis of ordinates, and x and z co-ordinates of a point m in the line $c'd'$, situated in the plane ZN, the equation of $c'd'$ is by (153)

$$x = az + \alpha.$$

In like manner the equation cd is

$$y = bz + \beta.$$

These equations determine the positions of two projections of CD, and hence (209) that of CD itself.

(211) *Schol.* 1. Of the four constants, a, α, b, and β, that enter into these equations, it is obvious that if none were known, nothing could be determined in regard to the position of the line to which the equations referred. If α only were known, it would fix the inclination of the plane $c'CDd'$ to the co-ordinate plane AF, but not its position, since an indefinite number of planes might be drawn parallel to it. If a and α only were known, the precise position of the plane $c'CDd'$ could be determined, but nothing in regard to the position of the line CD in the plane. If a, α, and b were known, they would fix the direction of the line CD in the plane; but still there might be an indefinite number of lines drawn parallel to it in the plane, which would satisfy the equations equally well. But lastly, if a, α, b, and β are all known, they limit the line to a single position, as already shown.

(211a) *Schol.* 2. Since a determines the inclination of the plane c'Cd', and b the direction of the line CD in that plane, the two together must determine the direction of any line in space; so that if these letters have the same value in the equation of any one line that they have in any other, the lines which the equations represent must be parallel, whatever may be the values of α and β.

(212) Prop. III. Theorem.

The equations of a straight line passing through a given point in space are

$$x'-x=a\ (z'-z),\quad \text{and}\quad y'-y=b\ (z'-z)\,;$$

in which x', y', and z' denote the co-ordinates of the given point, and the other letters as in the preceding proposition.

The same construction remaining as in the last proposition, let C be the given point, and c' and c its projections. Draw the ordinates $c\ t$ and ct.

Then $c't=\mathrm{C}c=x'$, $ct=\mathrm{C}c'=y'$, and by (203) $\mathrm{A}t=z'$.

Hence, in the same manner as in (156), the equations of the projections $c'd'$ and cd are

$$x'-x=a\ (z'-z),\quad \text{and}\quad y'-y=b\ (z'-z)\,;$$

which determine the position of the line in the same manner as in the last proposition.

(212a) A straight line passes through a point in space whose co-ordinates are

$$x'=7,\quad y'=8,\quad \text{and}\quad z'=9\,;$$

its projection upon the plane ZN crosses the axis of abscissas at an angle of 58°, and its projection upon ZY at an angle of 45°. Required the area of the triangles which the projections form with the co-ordinate axes (195a).

Ans. 17,1125 and ,5.

(213) Prop. IV. Theorem.

The equations of a straight line passing through two given points in space are

$$x'-x=\frac{x''-x'}{z''-z'}(z'-z), \quad \text{and} \quad y'-y=\frac{y''-y'}{z''-z'}(z'-z);$$

in which x'', y'', and z'' are the co-ordinates of one point, and x', y, and z' of the other.

In the same manner as in the two preceding propositions, it may be easily shown that the projections of the line upon the vertical planes, are straight lines passing through two given points in those planes, the co-ordinates of the two points being for the one projection (x'', z'') and (x', z'), and for the other projection (y'', z'') and (y', z'). Hence, by (157), the equation of one projection is

$$x'-x=\frac{x''-x'}{z''-z'}(z'-z);$$

and for the other $$y'-y=\frac{y''-y'}{z''-z'}(z'-z).$$

which together determine the position of the line in question.

(213*a*) *Ex.* A straight line passes through two points in space, whose co-ordinates are $x'=5$, $y'=5$, $z'=8$, and $x''=10$, $y''=4$, and $z''=6$. Required the points where its projections on the vertical planes cross the axis of abscissas, and at what angle.

Ans. One projection crosses above the origin at a distance of 10, and an angle of 116° 34′; and the other below the origin at a distance of 2, and an angle of 26° 34′.

(214) Prop. V. Theorem.

The distance between two points in space is

$$\sqrt{(x'-x)^2+(y'-y)^2+(z'-z)^2};$$

in which x', y', and z' are the co-ordinates of one point, and x, y, and z of the other.

If through each of the points three planes be made to pass, parallel to the co-ordinate planes, it is obvious that they will by their mutual intersection form a parallelopiped, of which the distance between the two points will be the diagonal, and whose edges will be the difference of the corresponding ordinates, viz.: $(x'-x)$, $(y'-y)$, and $(z'-z)$. But the square of the diagonal of a parallelopiped is equal to the sum of the square of its edges.[a] Hence, if we let D represent the distance between the points, we shall have

(214*a*) $$D^2=(x'-x)^2+(y'-y)^2+(z'-z)^2;$$

and extracting the root,

$$D=\sqrt{(x'-x)^2+(y'-y)^2+(z'-z)^2}.$$ [b]

(214*b*) *Ex.* Required the distance between two points in space; the co-ordinates of one being 8, 9, and 10, and of the other 3, 4, and 5. *Ans.* 8,66.

(215) Prop. VI. Theorem.

The tangent of the angle included between two straight lines in space is

$$\frac{\sqrt{(a'-a)^2+(b'-b)^2+(ab'-a'b)^2}}{1+aa'+bb'};$$

in which a and b represent the tangents of the angles that the projections of one of the lines make with the axis of abscissas, and a' and b' those of the other.

Let the equations of the two lines be

$$\begin{cases} x=az+\alpha, \text{ and } y=bz+\beta, \text{ for the one;} \\ x=a'z+\alpha', \text{ and } y=b'z+\beta', \text{ for the other.} \end{cases}$$

---

[a] Let AG be a parallelopiped, and DF its diagonal.

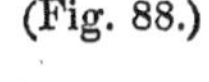
(Fig. 88.)

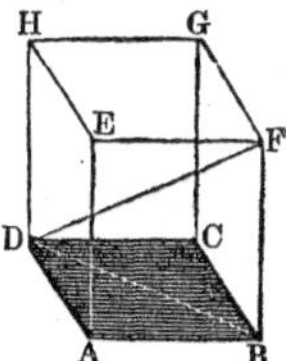

Because DAB is a right angle $DB^2=DA^2+AB^2$, and because DBF is a right angle $DF^2=DB^2+BF^2$; therefore $DF^2=DA^2+AB^2+BF^2$.

[b] This proposition may be illustrated by taking two points on the surface of an apple, and while the apple remains fixed in position, cutting it through each of the points in three directions parallel to three co-ordinate planes.

Then (211*a*) will the equations of two other lines, as AM and AN, drawn parallel to them through the origin, be

(215*a*) $\begin{cases} x=az, \quad \text{and } y=bz, \quad \text{for the one, AN;} \\ x=a'z, \quad \text{and } y=b'z, \quad \text{for the other, AM.} \end{cases}$

Now if from any point two lines be drawn parallel to any other two lines in space, the angle which the two latter make with each other is considered the same as that made by the two former, even though the latter do not lie in the same plane, so as to actually intersect.[a] Consequently, we have only to determine the angle formed by the two latter lines.

(Fig. 89.)

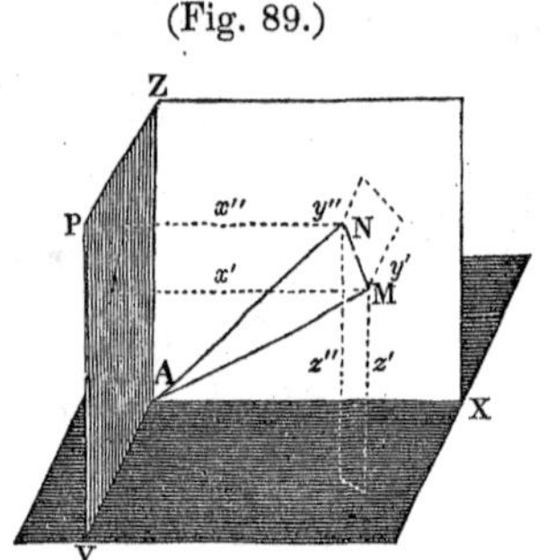

From any point N in AN draw NM perpendicular to AM, and denote the co-ordinates of M by x', y', and z', and those of N by x'', y'', and z''. And since the co-ordinates of A are zero, we have by (214) the distances AM, AN, and MN, as follows:

(215*b*) $$\begin{cases} \text{AM}=\sqrt{x'^2+y'^2+z'^2}; \\ \text{AN}=\sqrt{x''^2+y''^2+z''^2}; \\ \text{MN}=\sqrt{(x''-x')^2+(y''-y')^2+(z''-z')^2}. \end{cases}$$

The angle AMN being a right angle, we have

(215*c*) $$\text{AN}^2=\text{AM}^2+\text{MN}^2;$$

which, by transposing AM^2 and dividing by it, becomes

(215*d*) $$\frac{\text{AN}^2}{\text{AM}^2}-1=\frac{\text{MN}^2}{\text{AM}^2}.$$

Also, by trigonometry, we have

$$\tan.\ \text{MAN}=\frac{\text{MN}}{\text{AM}};$$

(215*e*) Or, $$\tan.^2\ \text{MAN}=\frac{\text{MN}^2}{\text{AM}^2}=(\text{by } 215d)\ \frac{\text{AN}^2}{\text{AM}^2}-1.$$

---

[a] Leg. 6. 6, Schol.

By substituting the values of AM, AN, and MN from (215*b*) into (215*c*) and (215*e*), and reducing the former, we obtain

$$(215f)\qquad x'^2+y'^2+z'^2=x''x'+y''y'+z''z'.$$

$$(215g)\qquad \tan.^2 \mathrm{MAN}=\frac{x''^2+y''^2+z''^2}{x'^2+y'^2+z'^2}-1.$$

As the equations (215*a*) are applicable to any points in the lines AM and AN, they will apply to the points M and N, and for these points they become

$$(215h)\qquad \begin{cases} x'=az', \text{ and } y'=bz', \text{ for AM.} \\ x''=a'z'', \text{ and } y''=b'z'', \text{ for AN.} \end{cases}$$

These values of x', y', x'', and y'' being substituted into (215*f*) and (215*g*), we have

$$(215k)\qquad (a^2+b^2+1)\,z'^2=(1+aa'+bb')\,z''z', \text{ or } \frac{z''}{z'}=\frac{a^2+b^2+1}{1+aa'+bb'}.$$

$$(215l)\qquad \tan.^2 \mathrm{MAN}=\frac{(a'^2+b'^2+1)}{a^2+b^2+1}\left(\frac{z''}{z'}\right)^2-1.$$

If now we substitute the value of $\frac{z''}{z'}$ from (215*k*) into (215*l*), we have, after cancelling the factor (a^2+b^2+1),

$$(215m)\qquad \tan.^2 \mathrm{MAN}=\frac{(a'^2+b'^2+1)\,(a^2+b^2+1)-(1+aa'+bb')^2}{(1+aa'+bb')^2};$$

which can be reduced to the form

$$\tan.^2 \mathrm{MAN}=\frac{(a'-a)^2+(b'-b)^2+(ab'-a'b)^2}{(1+aa'+bb')^2};$$

$$(215n)\qquad \text{Or, } \tan. \mathrm{MAN}=\pm\frac{\sqrt{(a'-a)^2+(b'-b)^2+(a'b-ab')^2}}{1+aa'+bb'};^{*}$$

---

* The formula given by other authors is

$$\cos. \mathrm{MAN}=\frac{1+aa'+bb'}{\sqrt{(a'^2+b'^2+1)\,(a^2+b^2+1)}};$$

but we have preferred the one we have adopted because it is equally simple, and preserves the analogy between straight lines in space and straight lines in a plane, as will be seen by comparing it with the formula given in (160).

the two values being the tangents of the two adjacent angles that the lines form with each other.

(215p) *Ex.* Required the angle included between two lines in space whose equations are

$$x=5z+7, \quad \text{and } y=3z+8, \quad \text{of the first;}$$
$$x=4z+10, \quad \text{and } y=2z+11, \quad \text{of the other.}$$

Ans. 5° 11′.

(216) *Schol.* 1. As in (161), the numerator of the last fraction must become zero when the lines are parallel, and the denominator zero when they are perpendicular to each other. But the numerator under the radical sign consists of three perfect squares, each of which must therefore be positive; so that the numerator can become zero only by each of the three terms of which it is composed becoming so. Hence the conditions of parallelism between two lines in space are $a =a$ and $b'=b$, (as already shown in another way in (211a),) and of perpendicularity

$$1+aa'+bb'=0.$$

(217) *Schol.* 2. It can be demonstrated, that if we represent the angles which one of the lines form with the co-ordinate axes by X, Y, and Z, and those of the other by X′, Y′, and Z′, we shall have

$$\cos. \text{MAN}=\cos. \text{X} \cos. \text{X}'+\cos. \text{Y} \cos. \text{Y}'+\cos. \text{Z} \cos. \text{Z}'.$$[a]

(218) Prop. VII. Problem.

Two lines in space, straight or curved, intersect: determine the point of intersection.

At the point of intersection the co-ordinates are common to both lines, and may be found by solving any three of the equations of the lines for x, y, and z.

---

[a] Davies' Analytical Geometry.

Ex. Let a straight line, whose equations are

$$x=az+\alpha, \quad \text{and} \quad y=bz+\beta,$$

intersect a curved line, the equations of whose projections are

$$x^2=p\,(a+z), \quad \text{and} \quad y=b'z+\beta',$$

which designate a parabola whose plane is perpendicular to the co-ordinate plane AYPZ (Fig. 89).

If we solve the first, second, and fourth equations for x, y, and z, we shall obtain for the co-ordinates of the point of intersection,

$$x=a\frac{\beta'-\beta}{b-b'}+\alpha, \quad y=\frac{\beta'b-\beta b'}{b-b'}, \quad \text{and} \quad z=\frac{\beta'-\beta}{b-b'}.$$

If a different selection had been made from the four equations the *form* of the answers would have been different, but the real *values* the same, as might be made to appear by eliminating x, y, and z from the four equations, and thus obtaining an equation between the constants.

CHAPTER IV.

OF PLANE SURFACES.

Def. The *traces* of a plane are the lines in which it cuts the co-ordinate planes.

(219) Prop. I. Theorem.

The equation of a plane is $z=\frac{x}{a}+\frac{y}{b}+c$,

in which a and b represent the tangents of the angles that the traces on the vertical planes make with the axis of abscissas, c the part of the axis of abscissas intercepted between the plane and the origin, and x, y, and z the co-ordinates of any point in the plane.

(Fig. 90.)

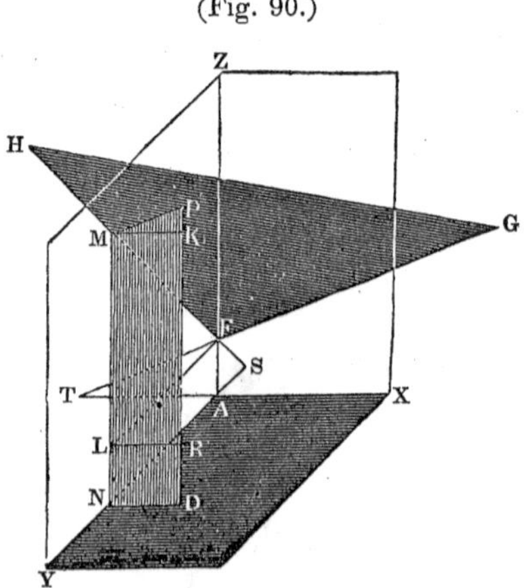

Let FGH be the plane, P any point in it, and FH and FG its traces on the vertical planes.

Through P draw the plane PMND parallel to ZX, meeting the trace FH in M, and the axis AY in N. Then will the lines ND, AN, and PD be equal to the co-ordinates of the point P; that is, ND$=x$, AN$=y$, and PD$=z$.

From F draw FL parallel to AY, and from L and M draw LR and MK parallel to AX

Then we shall have

$$MK=LR=ND=x, \quad FL=AN=y, \quad \text{and } RD=LN=AF=c.$$

Produce GF and HF till they meet the horizontal axes in T and S. Then, since the plane FGH cuts the two parallel planes MD and ZX, the lines of intersection MP and FG are parallel;[a] and consequently the angle PMK= GTX= the complement of GFZ. Also the angle HFL=HSY=the complement of HFZ.

Put tan. GFZ=a, and tan. HFZ=b.

$$\text{Then} \quad \tan. PMK=\cot. GFZ=(\text{by trigonometry})\ \frac{1}{a},$$

$$\text{and} \quad \tan. HFL=\cot. HFZ=\frac{1}{b}.$$

The ordinate PD, or z, consists of three parts PK, KR, and RD. Hence we have the equation

(219a) $$z=PK+KR+RD.$$

$$\text{But} \quad PK=\tan. PMK.MK=\frac{1}{a}x=\frac{x}{a}.$$

$$\text{And} \quad KR=ML=\tan. HFL.FL=\frac{1}{b}y=\frac{y}{b}.$$

$$\text{And} \quad RD=c.$$

Substituting these values into (219a), we have

$$z=\frac{x}{a}+\frac{y}{b}+c.$$

(219b) *Cor.* If the point be taken in the trace HF, the value of x becomes zero; and if in GF, the value of y becomes zero. Hence the equation of the trace HF is

$$z=\frac{y}{b}+c,$$

and of the trace GF, $$z=\frac{x}{a}+c.$$

The same thing may be shown also geometrically; for any ordinate $MN=ML+LN=\frac{y}{b}+c$; and in the same manner for the trace GF.

---

[a] Leg 6. 10. Euc. Sup. 2. 14.

(220) Prop. II. Theorem.

Every equation of the first degree between three variables is the equation of a plane.

For, by the ordinary operations of algebra, every such equation can be reduced to the form

(220*a*) $$Ax+By+Cz+D=0;$$

in which A, B, C, and D represent any known quantities whatever, whether positive or negative. But the above equation reduces to

(220*b*) $$z=-\frac{A}{C}x-\frac{B}{C}y-\frac{D}{C};$$

which corresponds to the form given in (219).

If now we measure off from the origin on the vertical axis a part equal to $-\frac{D}{C}$, and through the point thus found draw in the vertical planes two lines, so that the co-tangents of the angles which they make with the axis of abscissas shall be equal to $-\frac{A}{C}$, and $-\frac{B}{C}$, the plane of which these lines are the traces must have for its equation (220*a*) or (220*b*).

(220*c*) *Schol.* It may be shown, in the same manner as in (219*b*), that if the equation of a plane be given in the form (220*a*) the equations of its traces are

$$Ax+Cz+D=0, \text{ and } By+Cz+D=0.$$

(221) Prop. III. Problem.

To find the equation of a plane that shall pass through three given points.

Let the co-ordinates of the given points be (x', y', z'), (x'', y'', z''), (x''', y''', z'''). The equation of the plane must (219) be of the form

(221*a*) $$z=\frac{x}{a}+\frac{y}{b}+c.$$

As the equation must be true for every point in the plane, it must be true for the three given points. We shall therefore have the three following equations, viz.:

$$z' = \frac{x'}{a} + \frac{y'}{b} + c,$$

$$z'' = \frac{x''}{a} + \frac{y''}{b} + c,$$

$$z''' = \frac{x'''}{a} + \frac{y'''}{b} + c.$$

By means of these three equations we may obtain the values of the unknown quantities a, b, and c, which substituted into (221a) will give us the equation required. It can then be reduced to the following more simple form, viz.:

$$\frac{(x''-x)(z'-z)-(x'-x)(z''-z)}{(x''-x')(y'-y)-(x'-x)(y''-y')} = \frac{(x'''-x)(z'-z)-(x'-x)(z'''-z)}{(x'''-x')(y'-y)-(x'-x)(y'''-y')}.$$

(222) Prop. IV. Problem.

To find the equations of a straight line that shall be perpendicular to a given plane.

Let the equation of the given plane be

$$Ax + By + Cz + D = 0.$$

The plane that projects the line upon either of the co-ordinate planes, must be perpendicular both (204) to that co-ordinate plane and [a]to the given plane. Hence,[b] the projection of the line must be perpendicular to the trace of the plane.

Let the equations of the proposed line be

(222a) $$x = az + \alpha,$$

(222b) $$y = bz + \beta;$$

in which a, b, α, β are evidently unknown quantities, and that of the plane

$$Ax + By + Cz + D = 0.$$

---

[a] Leg. 6. 16. Euc. Sup. 2. 17. [b] Leg. 6. 18. Euc. Sup. 2. 18.

By (220c) the equations of the traces of the plane are

$$(222c)\qquad Ax+Cz+D=0,\quad \text{or } x=-\frac{C}{A}z-\frac{D}{A}.$$

$$(222d)\qquad By+Cz+D=0,\quad \text{or } y=-\frac{C}{A}y-\frac{D}{B}.$$

But the lines of which (222a) and (222c) are the equations lie in the same plane, and are perpendicular to each other. Hence we have, by (161),

$$(222e)\qquad 1+a\,(-\frac{C}{A})=0,\quad \text{or } a=\frac{A}{C};$$

and, in like manner, we have from (222b) and (222d),

$$(222f)\qquad 1+b\,(-\frac{C}{A})=0,\quad \text{or } b=\frac{B}{C}.$$

These values of a and b, substituted into (222a) and (222b), give the equations required, viz.:

$$(222g)\qquad x=\frac{A}{C}z+\alpha,\quad \text{and } y=\frac{B}{C}z+\beta.$$

(223) *Schol.* The values of α and β are left undetermined, which is as it should be, since the number of lines that can be drawn perpendicular to a plane is unlimited.

(224) Prop. V. Problem.

To find the inclination of a given line to a given plane.

(Fig. 92.)

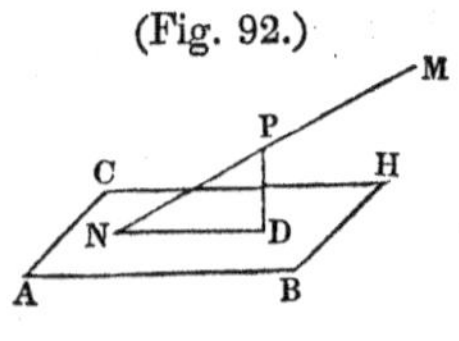

Let MN be the given line, having for its equations

$$x=az+\alpha,\quad \text{and } y=bz=\beta;$$

and let ABHC be the given plane, having for its equation

$$Ax+By+Cz+D=0.$$

From any point P in MN draw PD perpendicular to the plane, and join DN.

The equations of PD are, by (222g),

$$x=\frac{\mathrm{A}}{\mathrm{C}}z+\alpha, \text{ and } y=\frac{\mathrm{B}}{\mathrm{C}}z+\beta.$$

The inclination PND is the complement of the angle NPD.

Hence, by trigonometry, $\tan.\ \mathrm{PND}=\frac{1}{\tan.\ \mathrm{NPD}}$.

Now the equations of MN and PD being known, we may obtain an expression for tan. NPD by (215), viz.: by substituting $\frac{\mathrm{A}}{\mathrm{C}}$ in the place of a', and $\frac{\mathrm{B}}{\mathrm{C}}$ in the place of b'; and then, by inverting the fraction, we have its reciprocal, which is the tangent of the angle required. The following is the result, after multiplying both numerator and denominator by C,

$$\tan.\ \mathrm{PND}=\pm\frac{\mathrm{A}a+\mathrm{B}b+\mathrm{C}}{\mathrm{C}\sqrt{(\frac{\mathrm{A}}{\mathrm{C}}-a)^2+(\frac{\mathrm{B}}{\mathrm{C}}-b)^2+\left(\frac{\mathrm{A}b-\mathrm{B}a}{\mathrm{C}}\right)^2}}.$$

(225) *Schol.* The numerator of this fraction must become zero when the line is parallel to the plane, and the denominator zero when it is perpendicular. But the latter can happen, as was shown in (216), only when the separate terms become so. Hence, the conditions of parallelism between a line and plane, whose equation is of the form $\mathrm{A}x+\mathrm{B}y+\mathrm{C}z+\mathrm{D}=0$, is

$$\mathrm{A}a+\mathrm{B}b+\mathrm{C}=0;$$

and of perpendicularity,

$$\frac{\mathrm{A}}{\mathrm{C}}=a, \text{ and } \frac{\mathrm{B}}{\mathrm{C}}=b;$$

the latter being the same result that was obtained in Prop. IV.

(226) Prop. VI. Problem.

To determine the inclination of two given planes.

Let the equations of the two planes be

$$\mathrm{A}x+\mathrm{B}y+\mathrm{C}z+\mathrm{D}=0,$$
$$\mathrm{A}'x+\mathrm{B}'y+\mathrm{C}'z+\mathrm{D}'=0.$$

From any point in space let fall a perpendicular upon each plane. Then, by (222*g*), the equations of the perpendiculars will be

$$x=\frac{A}{C}z+\alpha, \text{ and } y=\frac{B}{C}z+\beta; \text{ the equations of the first.}$$

$$x=\frac{A'}{C'}z+\alpha', \text{ and } y=\frac{B'}{C'}z+\beta'; \text{ the equations of the second}$$

The angle included between these perpendiculars, which is the supplement of the inclination of the planes, may be found by substituting $\frac{A}{C}$, $\frac{B}{C}$, $\frac{A'}{C'}$, and $\frac{B'}{C'}$ in the place of a, b, a', and b' in (215).

Making the substitution and multiplying both numerator and denominator by C′C, we obtain for the tangent of the inclination, or of its supplement,

$$\frac{\pm\sqrt{(A'C-AC')^2+(B'C-BC')^2+(A'B-AB')^2}}{A'A+B'B+C'C}.$$

(226*a*) *Ex.* Determine the inclination of two planes whose equations are

$$2x+3y+4z-10=0;$$
$$3x-5y-2z+5=0.$$

Ans. 59° 12′.

(227) *Schol.* In the same manner as in the last proposition, we see that the conditions of parallelism between two planes are,

$$\frac{A'}{C'}=\frac{A}{C}, \text{ and } \frac{B'}{C'}=\frac{B}{C};$$

and of perpendicularity,

$$A'A+B'B+C'C=0.$$

(228) Prop. VII. Problem.

To determine the position of the foot of a perpendicular let fall from a given point in space upon a given plane.

Let the equation of the plane be

(228*a*) $$Ax+By+Cz+D=0.$$

Since the perpendicular passes through a given point, its equations must (212) be of the form

(228*b*) $x'-x=a\,(z'-z)$, and $y'-y=b\,(z'-z)$;

and since it is perpendicular to the given plane, we have, by (222*e*) and (222*f*),

$$\frac{\mathrm{A}}{\mathrm{C}}=a, \text{ and } \frac{\mathrm{B}}{\mathrm{C}}=b;$$

which values of a and b being substituted into (228*b*), we have

(228*c*) $$x'-x=\frac{\mathrm{A}}{\mathrm{C}}\,(z'-z);$$

(228*d*) $$y'-y=\frac{\mathrm{B}}{\mathrm{C}}\,(z'-z).$$

The three equations (228*a*), (228*c*), and (228*d*), solved for x, y, and z, make known the point required.

(228*e*) *Ex.* Find the position of the foot of a perpendicular let fall from a point in space whose co-ordinates are $x'=5$, $y'=6$, and $z'=7$, upon a plane whose equation is $4x+3y+2z+1=0$.

Ans. $$\begin{cases} x=-2\tfrac{9}{29}. \\ y=\tfrac{15}{29}. \\ z=3\tfrac{10}{29}. \end{cases}$$

(229) *Schol.* After determining the position of the foot of the perpendicular, its length may be found by (214)

CHAPTER V.

OF CURVED SURFACES.

---

(230) Prop. I. Theorem.

The equation of the surface of a sphere is

$$(m-x)^2+(n-y)^2+(p-z)^2=r^2;$$

in which r represents the radius of the sphere, m, n, and p the co-ordinates of the centre, and x, y, and z those of any point in the surface.

Since every point in the surface of the sphere is equally distant from the centre, the formula in (214*a*) for the distance between two points will apply to this case, one of the points being the centre of the sphere, and the other any point on the surface, and the distance between them the radius of the sphere. Therefore, by substituting r in the place of D in (214*a*), and m, n, and p in place of x', y', and z', we obtain the equation required, viz.:

$$(m-x)^2+(n-y)^2+(p-z)^2=r^2.$$

(231) *Cor.* If the origin is at the centre, m, n, and p disappear, and the equation reduces to $x^2+y^2+z^2=r^2$.

(232) Prop. II. Problem.

To find the equation of a plane tangent to a sphere.

If a plane touch a sphere, a straight line drawn from the centre to the point of contact is perpendicular to the plane. Consequently, if from any assumed point in the plane two lines be drawn, one to the centre of the sphere and the other to the point of tangency, these two lines, together with the radius of the sphere drawn to the point of tangency, will form a right-angled triangle.

Let the co-ordinates of the centre of the sphere be m, n, and p; those of the point of tangency x', y', and z'; and those of the assumed point in the plane x, y, and z.

By (214a) we may obtain the length of each side of the triangle in terms of these co-ordinates, viz.:

$\sqrt{(m-x)^2+(n-y)^2+(p-z)^2}=$ the distance from the assumed point to the centre of the sphere.

$\sqrt{(m-x')^2+(n-y')^2+(p-z')^2}=$ the radius of the sphere.

$\sqrt{(x'-x)^2+(y'-y)^2+(z'-z)^2}=$ the distance from the assumed point to the point of tangency.

Putting the square of the hypotenuse equal to the sum of the squares of the other two sides, we obtain

$$(232a)\quad (m-x)^2+(n-y)^2+(p-z)^2=(m-x')^2+(n-y')^2+(p-z')^2+(x'-x)^2+(y'-y)^2+(z'-z)^2;$$

which is the equation required.

(233) *Schol.* By expanding the terms in (232a) and transposing, it reduces to

$$(233a)\quad mx'-mx+x'x-x'^2+ny'-ny+y'y-y'^2+pz'-pz+z'z-z'^2=0;$$

which may be expressed in the form

$$(233b)\quad (m-x')(x'-x)+(n-y')(y'-y)+(p-z')(z'-z)=0.$$

Or,

$$(233c)\quad (x'-m)x+(y'-n)y+(z'-p)z+(m-x')x'+(n-y')y'+(p-z')z'=0.$$

If we add (233b) to the equation of the sphere, which may be expressed in the form

$$(m-x')(m-x')+(n-y')(n-y')+(p-z')(p-z')=r^2,$$

we have the equation of the tangent plane expressed in the form

$$(233d)\quad (m-x')(m-x)+(n-y')(n-y)+(p-z')(p-z)=r^2.$$

(233*e*) *Ex.* Determine the length of a perpendicular let fall from a point in space, whose co-ordinates x'', y'', and z'' are 5, 7, and 8 miles, upon a plane which is tangent to a sphere; the co-ordinates of the centre of the sphere being $m=4$ miles, $n=2$, and $p=5$; and of the point of tangency, $x'=3$ miles, $y'=1$, and $z'=6$.[a]

Ans. $\sqrt{12}$ miles.

(234). Prop. III. Theorem.

The equation of the surface of a right cone having a circular base is

$$(m-x)^2+(n-y)^2-a^2(p-z)^2=0;$$

in which m, n, and p represent the co-ordinates of the vertex of the cone; x, y, and z those of any point in the surface; and a the tangent of the angle that the generatrix (207) makes with the axis of the cone.

For convenience we will suppose the axis of the cone to be placed parallel to the axis of abscissas, so that any section parallel to the horizontal co-ordinate plane will be circular.

Through any point in the surface of the cone let a plane be made to pass parallel to the base. A circular section will thus be formed, the distance of whose centre from the vertex of the cone will be $p-z$; and, consequently, its radius will be $a(p-z)$. The horizontal co-ordinates of its centre will be the same as those of the vertex of the cone, viz.: m and n. We have, therefore, a circle having m and n for the horizontal co-ordinates of the centre, x and y for those of any point in the circumference, and $a(p-z)$ for its radius. Consequently (172) its equation is

$$(m-x)^2+(n-y)^2-a^2(p-z)^2=0;$$

and as this equation is true of every circle lying in the surface of the cone, it is the equation of the surface itself.

---

[a] The equation of the tangent plane in (233*c*) is the preferable form for the solution of this question, used in connection with (228) and (229).

(235) *Cor.* If the axis of a cone coincides with the vertical co-ordinate axis, m and n will disappear, and the equation will reduce to

$$x^2+y^2-a^2(p-z)^2=0;$$

and if the vertex be at the origin, it will reduce still farther to

$$x^2+y^2-a^2z^2=0.$$

(236) Prop. IV. Problem.

To find the equation of the surface of an ellipsoid.

As the generating ellipse (208) during its revolution constantly lies in the surface of the ellipsoid, it is evident that an equation that represents the former in every position, must represent the latter also.

(236a) Let $a^2y'^2+b^2x'^2-a^2b^2=0$ be the equation of the generating ellipse; m, n, and p the co-ordinates of the centre; and x, y, and z those of any point in the surface of the ellipsoid.

In this and the two succeeding propositions we will, for the sake of simplicity, suppose the axis of revolution to be parallel to the vertical co-ordinate axis, or axis of abscissas.

Whichever axis of the ellipse be taken as the axis of revolution, any point in the curve will describe a circle, the abscissa of whose centre will be z, the ordinates m and n, and its distance from the centre of the ellipse $z-p$. If it revolve about the transverse axis, this latter distance will also be x, and the radius of the circle y'; so that we have for the values of x' and y' the equations

(236b) $x'=z-p$; and, by (172), $y'^2=(x-m)^2+(y-n)^2$.

Substituting these values of x' and y' into (236a), we obtain for the equation

$$a^2(x-m)^2+a^2(y-n)^2+b^2(z-p)^2-a^2b^2=0.$$

which is the equation required, since it represents the ellipse in every position during its revolution.

The equation is found in the same way if the revolution be about the conjugate axis, only that x' and y' exchange places, so that the equation becomes

(236c) $$b^2(x-m)^2+b^2(y-n)^2+a^2(z-p)^2-a^2b^2=0.$$

(237) *Cor.* 1. If the axis of revolution coincides with the vertical co-ordinate axis, and the centre of the ellipse with the origin, m, n, and p will disappear, and the equations will become

(237a) $$\begin{cases} a^2(x^2+y^2)+b^2z^2-a^2b^2=0; \\ b^2(x^2+y^2)+a^2z^2-a^2b^2=0. \end{cases}$$

(238) *Cor.* 2. If $b=a$, (236c) becomes

$$(x-m)^2+(y-n)^2+(z-p)^2-a^2=0,$$

and (237a) becomes

$$x^2+y^2+z^2-a^2=0;$$

both of which are equations of the sphere.

(239) *Schol.* When the revolution is about the transverse axis, the ellipsoid is called a *prolate spheroid;* and when about the conjugate axis, an *oblate spheroid.*

(240) Prop. V. Problem.

To find the equation of the surface of a paraboloid.

As the generating parabola (208) during its revolution constantly lies in the surface of the paraboloid, it is evident that an equation that represents the former in every position, must represent the latter also.

(240a) Let $y'^2=p'x'$ be the equation of the generating parabola; m, n, and p the co-ordinates of its vertex; and x, y, and z those of any point in the surface of the paraboloid.

As the parabola revolves about its transverse axis, any point in the curve will describe a circle, whose radius will be y', the abscissa of its centre z, the ordinates m and n, and its distance from the vertex of the parabola $z-p$, and also x'. We have therefore the same equations as in (236b), and making the same substitution into (240a) that we did into (236a), we obtain the equation of the paraboloid, viz.:

$$(x-m)^2+(y-n)^2=p'\,(z-p).$$

(241) *Cor.* If the axis of revolution coincides with the vertical co-ordinate axis, and the vertex of the parabola with the origin, m, n, and p will disappear, and the equation will become

$$x^2+y^2=p'z.$$

(242) PROP. VI. PROBLEM.

To find the equation of the surface of an hyperboloid.

As the generating hyperbola (208) during its revolution constantly lies in the surface of the hyperboloid, it is evident that an equation that represents the former in every position, must represent the latter also.

Let $a^2y'^2-b^2x'^2+a^2b^2$ be the equation of the generating hyperbola; m, n, and p the co-ordinates of its centre; and x, y, and z those of any point in the surface of the hyperboloid.

It may be shown, by precisely the same process as in the ellipsoid (236), that if the revolution be about the transverse axis, the equation will be

$$\text{(242}a\text{)}\qquad a^2\,(x-m)^2+a^2\,(y-n)^2-b^2\,(z-p)^2+a^2b^2=0\,;$$

and if about the conjugate axis,

$$\text{(242}b\text{)}\qquad a^2\,(z-p)^2-b^2\,(x-m)^2-b^2\,(y-n)^2+a^2b^2=0.$$

(243) *Cor.* If we make the same supposition as in (237), equations (242a) and (242b) become

$$a^2\,(x^2+y^2)-b^2z^2+a^2b^2=0\,;$$
$$a^2z^2-b^2\,(x^2+y^2)+a^2b^2=0.$$

(244) *Schol.* 1. When the revolution is about the transverse axis, the hyperboloid consists of two parts separated from one

another, and it is called the *hyperboloid of two sheets;* but when about the conjugate axis, only a single solid is generated, which is called the *hyperboloid of one sheet.*

(245) *Schol.* 2. The process employed in the three preceding propositions will give us the equation of any solid of revolution, provided we know the equation of the generating curve.

(246) Prop. VII. Problem.

Two given surfaces, plane or curved, intersect: determine the line of intersection.

At the line of intersection the co-ordinates will be common to both surfaces. We may, therefore, by means of the equations of the two surfaces, eliminate one of the co-ordinates, and the resulting equation between two variables will be the equation of the line required.

Ex. 1. Let the surfaces be two planes, whose equations are

$$\mathrm{A}x+\mathrm{B}y+\mathrm{C}z+\mathrm{D}=0;$$
$$\mathrm{A}'x+\mathrm{B}'y+\mathrm{C}'z+\mathrm{D}'=0.$$

Eliminating z between these equations, we obtain

$$\left(\frac{\mathrm{A}}{\mathrm{C}}-\frac{\mathrm{A}'}{\mathrm{C}'}\right)x+\left(\frac{\mathrm{B}}{\mathrm{C}}-\frac{\mathrm{B}'}{\mathrm{C}'}\right)y+\left(\frac{\mathrm{D}}{\mathrm{C}}+\frac{\mathrm{D}'}{\mathrm{C}'}\right)=0, \text{ a straight line.}$$

Ex. 2. Let the equations of the surfaces be

$x^2+y^2-a^2z^2=0$, the surface of a cone,

$\mathrm{A}x+\mathrm{B}y+\mathrm{C}z+\mathrm{D}=0$, a plane.

Eliminating z and uniting terms, we obtain

$$\left(\mathrm{B}^2-\frac{\mathrm{C}^2}{a^2}\right)y^2+2\mathrm{AB}xy+\left(\mathrm{A}^2-\frac{\mathrm{C}^2}{a^2}\right)x^2+2\mathrm{BD}y+2\mathrm{AD}x+\mathrm{D}^2=0;$$

which is, by (193), the equation of a conic section, and we thus verify the results which we obtained geometrically at (57*a*), (78*a*), and (120*a*).

(247) *Schol.* Articles (192) and (193) enable us to determine, in any given case, whether the section is an ellipse, a parabola, an hyperbola, a circle, or a straight line.

Ex. A plane whose equation is $3x+2y+4z+1=0$, intersects a cone the equation of whose surface is $x^2+y^2-9z^2=0$; determine the line of intersection.

Ans. An hyperbola whose axes are ,682 and 534.

APPENDIX.

[*See page* 18.]

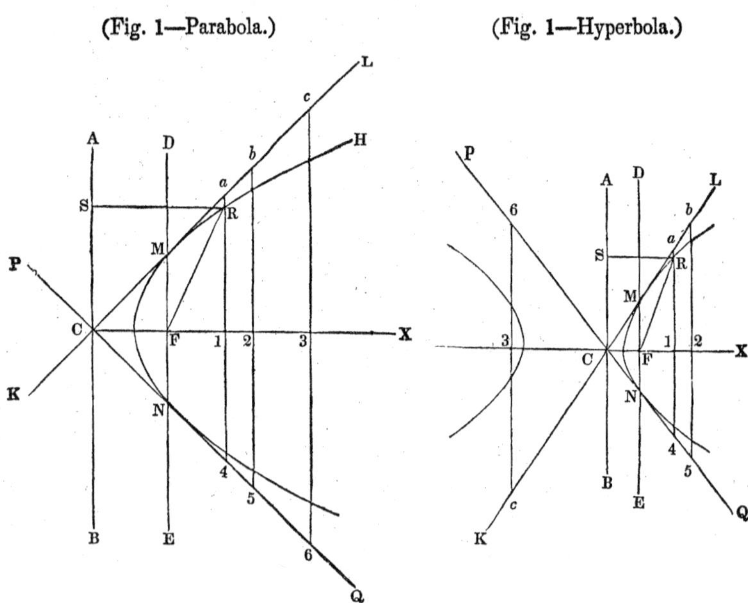

(Fig. 1—Parabola.) (Fig. 1—Hyperbola.)

A. [*See page* 13.]

It is proved in Bridge's treatise on the Conic Sections, that if a sphere be inscribed in a cone, so as to touch the plane of any conic section, the point of contact is the focus; and the line in which the plane of the conic section intersects that of the circle, formed by the mutual contact of the cone and the sphere, the directrix.

Let CDE represent a triangular section passing through the vertex of the cone, and perpendicular both to the plane of the base and the plane of the conic section, cutting the former in the line DE, the latter in AB, and the inscribed sphere in FGK. Then will the point F be the focus of the conic section, and the point H, in which the lines BA and KG meet when produced, will be in the directrix; and consequently, AF must be to AH in the given ratio mentioned in (1).

(Fig. 93.)

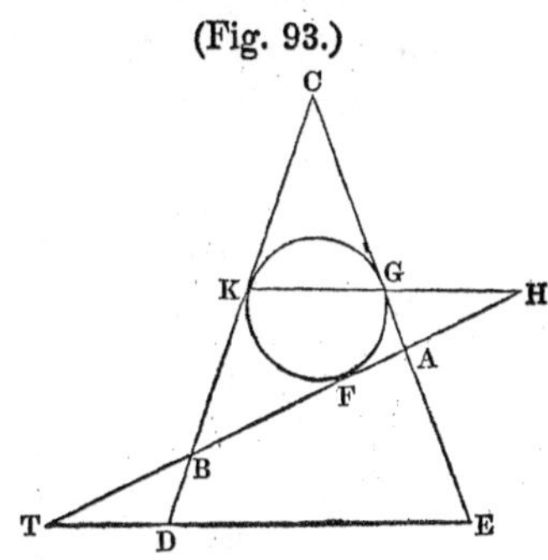

Produce AB and ED till they meet in T, forming a triangle ATE similar to AGH.

Then AG : AH :: AE : AT :: sin. ATE : sin. AET.

But AG=AF, both being tangents to a circle from the same point.

Therefore AF : AH :: sin. ATE : sin. AET, as remarked in (16*a*).

B. [*See page* 96.]

Another process, more strictly analytical, for finding the equation of a tangent to a circle (and the same general method will apply to any other plane curve), is the following:

Through P, the point of tangency, draw the line PP′ cutting the circle in any other point P′.

(Fig. 94.)

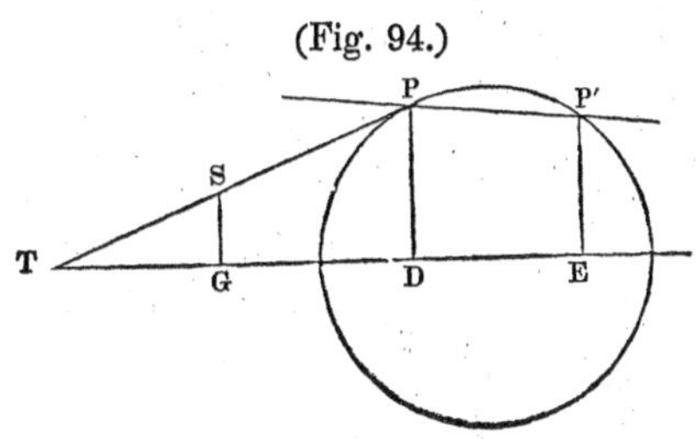

Since both PT and PP′ pass through the point P, their equations will be (156) of the form

(1) $$y'-y=a\,(x'-x),$$

(2) $$y'-y''=a'\,(x'-x'');$$

in which x' and y' designate the co-ordinates of the point P; x and y those of any point S in PT; x'' and y'' those of any point in PP′, (and consequently may represent those of P′); and a and a' the tangents of the angles which PT and PP′ respectively make with the axis of abscissas.

Further, since the points P and P′ are both in the circumference of the circle, we have (173) the equations

$$(3) \qquad x'^2+y'^2-r^2=0,$$

$$(4) \qquad x''^2+y''^2-r^2=0.$$

Subtracting (4) from (3), we obtain

$$x'^2-x''^2+y'^2-y''^2=0;$$

which resolved into factors reads

$$(x'+x'')(x'-x'')+(y'+y'')(y'-y'')=0.$$

Hence $y'-y''=-\dfrac{(x'+x'')(x'-x'')}{y'-y''}$.

Substituting this value of $y'-y''$ into (2), and then dividing both members by $x'-x''$, we obtain

$$(5) \qquad -\frac{x'+x''}{y'+y''}=a'.$$

If now we suppose the line PP′ to turn round the point P, P′ may be made to approach P; and when these points coincide, the line PP′ will coincide with PT, and we shall have

$$x'=x'', \quad y'=y'', \quad \text{and } a=a'.$$

Hence (5) will reduce to the form

$$-\frac{x'}{y'}=a.$$

Substituting this value of a into (1), we obtain the equation

$$y'-y=-\frac{x'}{y'}(x'-x);$$

which is the equation required, and agrees with that obtained at (175*b*).

C. [*See page* 109.]

It affords a fine illustration of the beauty of analytical processes of investigation, to observe the changes that the radius vector of an hyperbola undergoes as it revolves about the focus, causing ω to take different values from $0°$ to $360°$.

By (89a) $m^2-a^2=b^2$, so that the expressions for the value of r in (191c) may be read

$$r=\frac{b^2}{a-m\cos.\omega}, \text{ and } r=-\frac{b^2}{a+m\cos.\omega}.$$

If we make $\omega=0°$, we have $\cos.\omega=1$, and the expressions for r in (191c) become

$$r=-\frac{a^2-m^2}{a-m}=-a-m, \text{ and } r=a-m.$$

which are obviously the values of FA and FB (Fig. 27). Both being negative (since $m>a$), shows that the proper radius vector does not meet the curve, but that produced backward it meets it in two points, viz.: at A and B.

It is evident, moreover, that as the radius vector revolves both values of r continue negative, while $m\cos.\omega>a$; and therefore that up to that limit the radius vector produced backward meets both branches of the curve.

When $m\cos.\omega=a$, the two values of r become

$$r=-\frac{a^2-m^2}{0}=\infty, \text{ positive; and } r=\frac{a^2-m^2}{2a}=(89a)\ \frac{-b^2}{2a}=(87)\ -\frac{p}{4};$$

the former of which may easily be shown to be parallel to the asymptote, so that it does not meet the curve, and the latter meets it in the negative direction.

If we make $m\cos.\omega<a$, and $\cos.\omega$ positive, as it will be while $\omega<90°$, the first value of r becomes positive and the second negative, showing that the radius vector meets one branch of the curve in both directions, but does not meet the other at all.

If $\omega=90°$, we have cos. $\omega=0$, and the expressions become

$$r=\pm\frac{a^2-m^2}{a}=\text{(89}a\text{) and (87)}\pm\tfrac{1}{2}p.$$

as was shown in (183*b*).

The value of ω still increasing we shall have cos. ω negative, which will render the first value of r positive, and the second negative, so long as m cos. $\omega<a$, but infinite when m cos $\omega=a$. In the latter case the radius vector becomes parallel to the other asymptote.

If m cos. $\omega>a$, both values of r become positive, which shows that the proper radius vector meets both branches of the curve, but that produced backward it does not meet either.

When $\omega=180°$, we have cos. $\omega=-1$, and the expressions for r become

$$r=-\frac{a^2-m^2}{a+m}=-a+m,\quad\text{and}\quad r=\frac{a^2-m^2}{a-m}=a+m.$$

the same values as when $\omega=0°$, but with the opposite sign.

If we follow the radius vector through the two remaining quadrants, we shall find that the changes in the third quadrant correspond to those of the first, and those in the fourth to the second, but with the opposite signs.

D. Problem.

To double a given cube.[a]

Put a=the side of the given cube, and y=the side of the required cube.

Draw two parabolas having a common vertex, and their transverse axes at right angles to one another, the parameter of one being equal to a, and that of the other to $2a$. From the point where the parabolas intersect draw an ordinate to the axis of that which has

[a] This was a problem of great celebrity among the ancients.

the greatest parameter, and the abscissa will be the side of the required cube.

For at the point of intersection the co-ordinates are common to both curves, but the abscissa of one is the ordinate of the other, and *vice versa*. We have, therefore, the two equations

$$y^2=ax,$$
$$x^2=2ay.$$

Eliminating x between these equations and reducing, we obtain

$$y^3=2a^3.$$

Cor. If we eliminate y instead of x, we shall obtain

$$x^3=4a^3;$$

so that the same construction enables us to quadruple a given cube.

E. (*See page* 11.)

Two hyperbolas so drawn that the transverse axis of one is the conjugate axis of the other, and *vice versa*, are called *conjugate hyperbolas;* and such hyperbolas only have conjugate diameters.

F. (*See page* 25.)

Instead of the enunciation and demonstration employed in this proposition, a form analogous to Art. 104 may be substituted, if thought preferable.

ROBERT B. COLLINS,

BOOKSELLER AND STATIONER, 254 PEARL-STREET, N. Y.

Invites the attention of those interested to the following valuable books, which, it is believed, will be found on examination *unsurpassed* as text books.

MATHEMATICS.—ADAMS'S ARITHMETICAL SERIES.

I. Primary Arithmetic.—An introduction to Adams's New Arithmetic, revised edition.
II. Adams's New Arithmetic, revised edition; being a revision of Adams's New Arithmetic, first published in 1827. 12mo. half bound.
A Key to the above is published separately.
III. Mensuration, Mechanical Powers, and Machinery, a sequel to the Arithmetic.
IV. Bookkeeping by Single Entry, accompanied with blank books for the use of learners.

Abbott's Arithmetic.

The Mount Vernon Arithmetic.—Part I., Elementary. By Jacob and Charles E. Abbott.
The Mount Vernon Arithmetic.—Part II., Vulgar and Decimal Fractions. By Jacob Abbott. 12mo., half bound.

McCurdy's Geometry.

First Lessons in Geometry. By D. McCurdy.
Charts to accompany the "First Lessons," on rollers, size 34 by 48 inches.
Euclid's Elements, or Second Lessons in Geometry. By D. McCurdy. 12mo. half bound.

Preston's Bookkeeping.

District School Bookkeeping. Quarto, stitched. An excellent work for beginners; printed on handsome demy paper for practice.
Single Entry Bookkeeping. 8vo. cloth sides.
Bookkeeping, by single and double entry.
These three excellent works are by Lyman Preston, the author of the Interest Tables, &c.

Coffin's Conic Sections.

Elements of Conic Sections and Analytical Geometry. By James H. Coffin, Prof. Mathematics, Lafayette College, Pa.

Day's Mathematics.

A Course of Mathematics, containing the principles of Plane Trigonometry, Mensuration, Navigation, and Surveying. By Jeremiah Day, D. D. LL. D., President of Yale Col. 8vo. sheep.

PHILOSOPHY AND ASTRONOMY.

Olmsted's Rudiments of Natural Philosophy. 18mo.
Olmsted's Rudiments of Astronomy. 18mo.
These two works are intended to be used in district and common schools, where merely the elements of the sciences are taught, and also as preparatory class-books for the younger scholars in academies; for these purposes they are admirably adapted. They are in a neat 18mo. form, well bound in cloth, with leather back, and are illustrated by numerous engravings.
Olmsted's School Astronomy. 12mo.
A compendium of Astronomy for the use of pupils in the higher classes of common schools and in academies.
Olmsted's College Astronomy. 8vo., sheep.
Olmsted's College Philosophy. 8vo., sheep.
These works were prepared by Prof. Olmsted for the use of his classes in Yale College, and are acknowledged to be the best works on the subject yet before the public; the author's name is, in itself, a sufficient guarantee of their excellence.
Mason's Practical Astronomy. 8vo. cloth, leather back.
This work, by the late E. P. Mason, is regarded by competent judges as the best work extant on the subject of Practical Astronomy.
Solar and Lunar Eclipses, familiarly illustrated and explained, with the method of calculating them, as taught in the New England Colleges. By James H. Coffin, A. M. 8vo.

MISCELLANEOUS.

The Mount Vernon Reader, in three numbers. By the Brothers Abbott
Addick's Elements of the French Language. 12mo., half cloth.
Girard's Elements of the Spanish Language.
Two excellent rudimentary works.
Badlam's Common School Writing Books.—A series of five numbers.
The Governmental Instructor.—A brief and comprehensive view of the government of the United States and of the state governments, for the use of schools. By J. B. Shurtleff.
Kirkham's English Grammar.—This work is more extensively used and approved than any work of its size in the country.
Whelpley's Compend of History, revised, with additional matter, by Joseph Emerson.
John Rubens Smith's Drawing Book. Folio.
The Oxford Drawing Book. 100 lithographic plates, with directions, &c. Quarto.
Æsop's Fables.—The most splendid edition ever published, with illustrations on wood, by Orr, Howland, and Felter, after designs by John Tenniel. Crown 8vo., fancy cloth.
Gabriel, a story of Wichnor Wood. By Mary Howitt. The best tale of this popular authoress.
Our Cousins in Ohio, a picture of domestic life in the West. By Mary Howitt. 18mo., cloth gilt.
The Imitation of Christ. By T. à Kempis. Payne's translation, with essay by Dr. Chalmers. The cheapest edition of this valuable work ever issued from the press.
Abercrombie's Intellectual Powers.—Inquiries concerning the Intellectual Powers, and Investigation of Truth. By John Abercrombie. 12mo., cloth.
Abercrombie's Moral Philosophy.
These two works of Abercrombie's are made much more valuable by the additions and questions to this edition by Jacob Abbott.
Dymond's Essays on the Principles of Morality, and the private and political rights and obligations of Mankind. This is regarded as the best work on Ethics in the language. 12mo.

www.ingramcontent.com/pod-product-compliance
Lightning Source LLC
LaVergne TN
LVHW021402110826
845150LV00007B/1758

* 9 7 8 1 4 2 5 5 1 2 6 5 1 *